Lecture Notes in Computer Scie

T0238020

Commenced Publication in 1973
Founding and Former Series Editors:
Gerhard Goos, Juris Hartmanis, and Jan van Leeuwen

Anthony Bonato   Fan R.K. Chung (Eds.)

Algorithms and Models for the Web-Graph

5th International Workshop, WAW 2007
San Diego, CA, USA, December 11-12, 2007
Proceedings

 Springer

Volume Editors

Anthony Bonato
Wilfrid Laurier University, Department of Mathematics
Waterloo, ON, N2L 3C5, Canada
E-mail: abonato@rogers.com

Fan R.K. Chung
University of California, San Diego, Department of Mathematics
La Jolla, CA 92093-0112, USA
E-mail: fan@math.ucsd.edu

Library of Congress Control Number: 2007939821

CR Subject Classification (1998): F.2, G.2, H.4, H.3, C.2, H.2.8, E.1

LNCS Sublibrary: SL 1 – Theoretical Computer Science and General Issues

ISSN   0302-9743
ISBN-10   3-540-77003-8 Springer Berlin Heidelberg New York
ISBN-13   978-3-540-77003-9 Springer Berlin Heidelberg New York

Springer is a part of Springer Science+Business Media

springer.com

© Springer-Verlag Berlin Heidelberg 2007
Printed in Germany

Typesetting: Camera-ready by author, data conversion by Scientific Publishing Services, Chennai, India
Printed on acid-free paper   SPIN: 12197316   06/3180   5 4 3 2 1 0

Preface

This volume constitutes the refereed proceedings of the Fifth Workshop on Algorithms and Models for the Web-Graph, WAW 2007, held in San Diego in December 2007. The proceedings consist of 18 revised papers (13 regular papers and 5 short papers) which were reviewed and selected from a large pool of submissions. The papers address a wide variety of topics related to the study of the Web-graph such as random graph models for the Web-graph, PageRank analysis and computation, decentralized search, local partitioning algorithms, and traceroute sampling.

The Web-graph has been the focal point of a tremendous amount of research for more than a decade. The view of the Web as a graph has great practical importance and has also generated much interesting theoretical work. A goal of the 2007 Workshop was to present state-of-the art research on both the applications and theory of the Web-graph. Our hope is that the papers presented here will help stimulate new and exciting avenues of research on the Web-graph.

December 2007

Anthony Bonato
Fan Chung Graham

Organization

Executive Committee

Conference Chair	Ronald Graham (University of California, San Diego, USA)
Local Arrangements Chair	Tara Javidi (University of California, San Diego, USA)
Program Committee Co-chair	Anthony Bonato (Wilfrid Laurier University, Canada)
Program Committee Co-chair	Fan Chung Graham (University of California, San Diego, USA)
Program Committee Co-chair	Tara Javidi (University of California, San Diego, USA)

Organizing Committee

Andrei Broder, (Yahoo! Research, USA)
Fan Chung Graham (University of California, San Diego, USA)
Jeannette Janssen, (Dalhousie University, Canada)
Tara Javidi (University of California, San Diego, USA)
Lincoln Lu (University of South Carolina, USA)

Program Committee

Dimitris Achlioptas, (University of California, Santa Cruz, USA)
Colin Cooper, (King's College London, UK)
Anthony Bonato (Wilfrid Laurier University, Canada)
Alan Frieze (Carnegie Mellon University, USA)
Michael Goodrich, (University of California, Irvine, USA)
Fan Chung Graham (University of California, San Diego, USA)
Jeannette Janssen, (Dalhousie University, Canada)
Tara Javidi (University of California, San Diego, USA)
Ravi Kumar (Yahoo! Research, USA)
Kevin Lang, (Yahoo! Research, USA)
Stefano Leonardi (Università di Roma, Italy)
Lincoln Lu (University of South Carolina, USA)
Milena Mihail (Georgia Institute of Technology, USA)
Michael Mitzenmacher (Harvard University, USA)
Muthu Muthukrishnan (Rutgers University and Google Inc., USA)
Joel Spencer (New York University, USA)
Walter Willinger (AT&T Research, USA)

Sponsoring Institutions

California Institute for Telecommunications and Information Technology
Google Inc.
Yahoo! Research
National Science Foundation
Springer *Lecture Notes in Computer Science*
University of California, San Diego

Table of Contents

Bias Reduction in Traceroute Sampling – Towards a More Accurate Map of the Internet

Abraham D. Flaxman[1] and Juan Vera[2]

[1] Microsoft Research
Redmond, WA
abie@microsoft.com
[2] Georgia Institute of Technology
Atlanta, GA
jvera@cc.gatech.edu

Abstract. Traceroute sampling is an important technique in exploring the internet router graph and the autonomous system graph. Although it is one of the primary techniques used in calculating statistics about the internet, it can introduce bias that corrupts these estimates. This paper reports on a theoretical and experimental investigation of a new technique to reduce the bias of traceroute sampling when estimating the degree distribution. We develop a new estimator for the degree of a node in a traceroute-sampled graph; validate the estimator theoretically in Erdős-Rényi graphs and, through computer experiments, for a wider range of graphs; and apply it to produce a new picture of the degree distribution of the autonomous system graph.

1 Introduction

The internet is quite a mysterious network. It is a huge and complex tangle of routers, wired together by millions of edges. To understand this *router graph* is quite a challenge, one that has driven research for the last decade.

The router graph has a natural clustering into Autonomous Systems (ASes), which are sets of routers under the same management. Producing an accurate picture of the *AS graph* is an important step towards understanding the internet.

There are three techniques for finding large sets of edges in the AS graph: the WHOIS database, BGP tables, and traceroute sampling. No approach is clearly superior, and the results of the different approaches are compared in detail in a recent paper [14].

The present paper focuses on traceroute sampling, an approach applicable to the router graph as well as the AS graph. Traceroute sampling consists of recording the paths that packets follow when they are sent from monitor nodes to target nodes, and merging all of these paths to produce an approximation of the AS graph.

A seminal analysis using both traceroute sampling and BGP tables concluded that the AS graph degree distribution follows a power-law (meaning that the number of ASes of degree k is proportional to $k^{-\alpha}$ for a wide range of k values)

A. Bonato and F.R.K. Chung (Eds.): WAW 2007, LNCS 4863, pp. 1–15, 2007.

[7]. This caused a shift in simulation methodology for evaluating network algorithms and also contributed to the avalanche of recently developed network models which produce power-law degree distributions.

However, the true nature of the AS-graph degree distribution was called into question by computer experiments on synthetic graphs [12,17]. These experiments show that if the sets of monitor and target nodes are too small then traceroute sampling will produce a power-law degree distribution, even when the underlying graph has a tightly concentrated degree distribution. Theoretical follow-up work proved rigorously that in many non-power-law graphs, including random regular graphs, an idealized model of traceroute sampling yields power-law degree distributions [4,1].

Subsequent computer experiments have led some to believe that the bias inherent to traceroute sampling can be ignored, at least for making a qualitative distinction between scale-free and homogeneous graphs, when using a large enough set of monitor nodes [9]. This is also supported by an analysis using the statistical physics technique of mean field approximation [5].

1.1 Our Contribution

This paper proposes a new way forward in the struggle to characterize the degree distribution of the AS graph. Our contribution has three parts:

1. We derive a statistical technique for reducing the bias in traceroute sampling;
2. We verify the technique experimentally and theoretically, in the framework previously studied in [12,4];
3. We use the traceroute bias-reduction technique to generate a more accurate picture of the AS degree distribution over time, which suggests that aspects of commercially available technology are reflected in the network topology.

Our approach for reducing the bias in traceroute sampling is based on a technique from biostatistics, the multiple-recapture census, which has been developed for estimating the size of an animal population [18] (this technique also has applications to proofreading [19]). However, we do not have the benefit of independent random variables which are central to the animal counting and proofreading statistics, and so we must adapt the technique to apply to random variables with complicated dependencies.

To provide some evidence that this bias-reduction technique actually reduces bias, we consider a widely used model of traceroute sampling, which assumes that data travels from monitor to target along the shortest path in the network. It is generally believed that the path that data actually takes is *not* the shortest path, but that the shortest path is an acceptable approximation of the actual path (see [13] for a discussion of this approximation). In this model, it is possible to check theoretically and experimentally that the bias reduction provides a better estimate of the degree distribution. We show that the new estimation is asymptotically unbiased for the Erdős-Rényi random graph $G_{n,p}$ when $np \gg \log n$, and that it gives improved estimates for finite instances from a variety of different graphs.

Finally, we use the bias-reduction technique on real data, traceroute samples from the internet. The new estimate of the AS-graph degree distribution is still scale-free over two orders of magnitude, with an exponent very similar to the uncorrected degree distribution (see Figure 1). A by-product of bias reduction is the removal of all vertices with degree less than 3, and this increases the average degree. For example, in March 2004 (the month used for comparison in [14]), the biased estimate of average degree is 6.29, while after bias reduction the average degree is 12.66 (which is very close to 12.52, the biased average degree when restricted to vertices of degree at least 3). An interesting feature in the bias-reduced AS degree distribution (from March 2004) is the lack of nodes with degree between 65 and 90; at the time, a popular router maker offered a router which provided up to 64 ports per chassis. In March 2002, before this product was available, there was no dearth of 65 degree nodes.

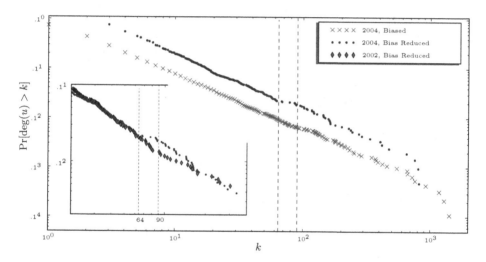

Fig. 1. Degree sequence ccdf estimates for the AS graph (from CAIDA skitter). Main panel: March, 2004, with and without bias reduction. Inset: a portion of ccdf for March, 2004 and March, 2002, both with bias reduction. The nodes with degree between 65 and 90 in 2002 have disappeared in 2004.

1.2   Related Work

Internet mapping by traceroute sampling was pioneered by Pansiot and Grad in [15], and the scale-free nature of the degree distribution was observed by Faloutsos, Faloutsos, and Faloutsos in [7]. Since 1998, the Cooperative Association for Internet Data Analysis (CAIDA) project *skitter* has archived traceroute data that is collected daily [10]. The bias introduced by traceroute sampling was identified in computer experiments by Lakhina, Byers, Crovella, and Xie in [12] and Petermann and De Los Rios [17], and formally proven to hold in a model of one-monitor, all-target traceroute sample by Clauset and Moore

[4] and, in further generality, by Achlioptas, Clauset, Kempe, and Moore [1]. Computer experiments by to Guillaume, Latapy, and Magoni [9] and an analysis using the mean field approximation of statistical physics due to Dall'Asta, Alvarez-Hamelin, Barrat, Vázquez, and Vespignani [5] argue that, despite the bias introduced by traceroute sampling, some sort of scale-free behavior can be inferred from the union of traceroute-sampled paths.

The present paper provides a new avenue for investigating these controversial questions, by developing a method for *correcting* the bias introduced by traceroute sampling. Another recent paper by Viger, Barrat, Dall'Asta, Zhang and Kolaczyk applied techniques from statistics to reduce the bias of traceroute sampling [21]. That paper focused on estimating the number of nodes in the AS graph, and applied techniques from a different problem in biostatistics, estimating the number of species in a bioregion. The problem of correcting bias in sampled networks has a long history in sociology, although the biases in that domain seem somewhat different; see the surveys by Frank, by Klovdahl, or by Salganik and Heckathorn for an overview [8,11,20].

In addition to traceroute sampling, maps of the AS graph have been generated in two different ways, using BGP tables and using the WHOIS database. A recent paper by Mahadevan, Krioukov, Fomenkov, Dimitropoulos, claffy, and Vahdat provides a detailed comparison of the graphs that result from each of these measurement techniques [14].

1.3   Outline of What Follows

The new estimator for the degree of a node in the AS graph is developed from multiple-recapture population estimation in Section 2. Section 3 argues that this estimator generates an asymptotically unbiased degree distribution for the Erdős-Rényi graph $G_{n,p}$ when $p \gg \log n$, which rigorously demonstrates that the new estimator can reject a null hypothesis. Section 4 presents additional evidence that the new estimator reduces the bias of traceroute sampling, in the form of computer experiments on synthetic networks. Section 5 provides a comparison between the degree sequence predicted by the new estimator and the previous technique, and details how, after bias reduction, the degree distribution may reflect economic and technological factors present in the system, i.e., there a significantly larger marginal cost of adding a 65th neighbor than adding a 64th neighbor when using the Juniper T320 edge router. Section 6 provides a conclusion and focuses on directions of future research to strengthen this approach.

2   Estimation Technique

The classical capture-recapture approach to estimating an animal population has two phases. First, an experimenter captures animals for a given time period, marks them (with tags or bands), and releases them, recording the total number of animals captured. Then, the experimenter captures animals for a second time period, and records both the number of animals recaptured and the total number of animals captured during the second period. If A denotes the number of animals captured in phase one, B denotes the number captured during phase two, and C denotes the number captured in phase one and captured again in phase two, then an estimate of total population size is given by

$$\widehat{N} = \begin{cases} \frac{AB}{C}, & \text{if } C \neq 0; \\ \infty, & \text{otherwise.} \end{cases}$$

If the true population size is N, and each animal is captured or not captured during each phase independently, with probability p_1 during phase one and probability p_2 during phase two, then $\widehat{N}$ is the maximum likelihood estimate of N [18]. For more than two phases, the maximum likelihood estimator does not have a simple closed form, but it can be computed efficiently using the techniques developed in [18].

When estimating the degree of a particular AS by traceroute sampling, each edge corresponds to an animal, and each monitor node corresponds to a recapture phase. Unfortunately, in this setting there is no reason to believe that the events "monitor i observes edge j" are independent. Indeed, when shortest-path routing is used (as an approximation of BGP routing), these events are highly dependent. However, it is still possible adapt the capture-recapture estimate to reduce bias in this case.

Let G be a graph, and let s and t be monitor nodes in G. Let G_s be the union of all routes discovered when sending packets from s to every node in the target set. Define G_t analogously. Let $N_s(u)$ denote the neighbors of u in G_s and define $N_t(u)$ analogously.

Using this notation, the modification of the capture-recapture estimate proposed for traceroute sampling is given by

$$\widehat{\deg}_{s,t}(u) = \begin{cases} \frac{|N_s(u)| \cdot |N_t(u)|}{|N_s(u) \cap N_t(u)|}, & \text{if } |N_s(u) \cap N_t(u)| > 2; \\ \infty, & \text{otherwise.} \end{cases}$$

When more than 2 monitor nodes are available, pair up the monitors, consider the estimates given by each pair that are not ∞, and for the final estimator, use the median of these values. So, if the monitor nodes are paired up as $(s_1, t_1), (s_2, t_2), ..., (s_k, t_k)$ then

$$\widehat{\deg}(u) = \text{median}\left(\left\{\widehat{\deg}_{s_i,t_i}(u) \neq \infty\right\}\right).$$

This degree estimator can also provide an estimate of the cdf of the degree distribution (i.e., the fraction of nodes with degree at most k) according to the formula

$$\widehat{d_{\leq k}} = \widehat{\Pr}[\deg(u) \leq k] = \frac{\#\{u : \widehat{\deg}_{s,t}(u) \leq k\}}{\#\{u : \widehat{\deg}_{s,t}(u) < \infty\}}.$$

Discussion: It may seem wasteful to consider the median-of-two-monitors estimate instead of combining all available monitors in a more holistic manner.

However, we have conducted computer experiments with maximum likelihood estimators for multiple-recapture population estimation with more than two phases, and the adaptations we have considered thus far perform significantly worse than the median-of-two-monitors approach above. This is probably due to the complicated dependencies of several overlapping shortest-path trees. However, the exploration we have conducted to date is not exhaustive, and does not rule out the possibility that a significantly better estimator exists.

3   Theoretical Analysis

This section and the next intend to provide some assurance that repeated application of $\widehat{\deg}(u)$ is an accurate way to estimate the degree distribution of the sampled graph.

This section provides a theoretical analysis of the performance of $\widehat{\deg}(u)$ in a very specific setting: when the underlying graph is the Erdős-Rényi graph $G_{n,p}$ with n sufficiently large, $np \gg \log n$, and every vertex is a target node. For the purpose of analysis, this section and the next assume that traceroute finds a shortest path from monitor to target. This is the same setting that is considered in [4].

Theorem 1. *Let $G \sim G_{n,p}$ be a random graph with $np = d \gg \log n$, and let s, t, and u be uniformly random vertices of G. Then, for any k, with high probability,*

$$\widehat{d_{\leq k}} = \frac{\#\{u : \widehat{\deg}(u) \leq k\}}{\#\{u : \widehat{\deg}(u) < \infty\}} = \frac{\#\{u : \deg(u) \leq k\}}{n} \pm \mathcal{O}\left(1/d\right).$$

Proof sketch: The analysis *two* breadth-first-search trees in a random graph is difficult when the average degree is small. But, for d moderately large, as in this theorem, the situation is simpler.

It follows from the branching-process approximation of breadth-first search that with high probability there are $(1 \pm \epsilon)d^i$ vertices at distance exactly i from s (or t) when $i < (\log n)/(\log d)$. Thus, almost all vertices are distance $\lceil (\log n)/(\log d) \rceil$ apart. For ease of analysis, suppose that $\ell = (\log n)/(\log d)$ is an integer.

So, with high probability, if u is at distance ℓ from s or t then it is a leaf node in G_s or G_t. In this case, $|N_s(u) \cap N_t(u)| \leq 1$ and therefore $\widehat{\deg}(u) = \infty$.

Now, consider the case where vertex u is distance i from s and distance j from t, where $i, j < \ell$. Let $N(u)$ denote the neighbors of u in G, and then let S be the set of vertices within distance i of s in G and let T be the set of vertices within distance j of t in G. Conditioned on S, T and $N(u)$, the set of indicator random variables

$$\left\{ \mathbf{1}[v \in N_s(u)], \mathbf{1}[v \in N_t(u)] : v \in N(u) \setminus (S \cup T) \right\}$$

is independent, and, for $v \in N(u) \setminus (S \cup T)$, $\Pr[v \in N_s(u)]$ and $\Pr[v \in N_t(u)]$ are functions of S and T, but constants with respect to v, i.e., $\Pr[v \in N_s] = p_s$

and $\Pr[v \in N_t] = p_t$. So, besides any edges between u and $S \cup T$, the edges incident to u in $G_s[S]$ and $G_t[T]$ yield the random variables $|N_s(u)|$, $|N_t(u)|$, and $|N_s(u) \cap N_t(u)|$, which correspond to A, B, and C in the capture-recapture estimate of population size. For example, if there is only one edge incident to u in $G_s[S]$ and only one in $G_t[T]$, and these edges are different, then

$$\Pr\left[\widehat{\deg(u)} \geq k \mid S, T, N(u)\right] = \Pr\left[\frac{(A+1)(B+1)}{C} \geq k\right],$$

where $C \sim \mathrm{B}(|N(u)| - 2, p_s p_t)$, $A \sim C + \mathrm{B}(|N(u)| - 1 - C, p_s)$, and $B \sim C + \mathrm{B}(|N(u)| - 1 - C, p_t)$. If k is sufficiently large and p_s and p_t are not too small then this probability is concentrated in the range $k = |N(u)| \pm \sqrt{|N(u)|}$.

To complete the proof, it remains to show that, with probability $1 - \mathcal{O}(1/d)$, $p_s, p_t \geq \epsilon$ and $|N(u) \cap (S \cup T)| \leq 2$, and from this show that, for A, B, C defined analogously to above,

$$\Pr\left[\frac{(A+1)(B+1)}{C} \geq k\right] = \Pr[|N(u)| \geq k] + \mathcal{O}(1/d). \qquad \square$$

Discussion: This analysis would go through without modification if the estimate also included samples where $|N_s(u) \cap N_t(u)| = 2$, but the definition of $\widehat{\deg}(u)$ from above seems to behave better under finite scaling.

The proof sketch can be adapted for random graphs with other degree distributions, provided that the average degree is large. However, the proof relies on the fact that the graph is *locally tree-like*, which ensures that $N(u) \cap (S \cup T)$ is likely to be small. This assumption does not seem to hold in the AS graph, and even G_s, the union of all routes discovered from a single monitor node s, has some triangles. The next section includes evidence from computer experiments that in graphs which are *not* locally tree-like, such as the random geometric graph, estimator $\widehat{\deg}(u)$ is not asymptotically unbiased, but can still reduce some amount of bias. Proving this rigorously may be a difficult task.

4   Computer Experiments

This section describes the results of a series of computer experiments conducted to investigate how well $\widehat{d_{\leq k}}$ approximates the true degree distribution.

We consider three different distributions for random graphs, the Erdős-Rényi model, the Preferential Attachment model, and the random geometric graph. Additionally, we consider synthetic data based on a real-world graph, the Western States Power Grid (WSPG), which Duncan Watts has graciously made available to researchers [22]. These graphs will all be described in more detail below.

For each graph, we set edge e to be of length $1 + \eta_e$, where η_e is selected uniformly from the interval $[-1/n, 1/n]$, where n is the number of vertices. This ensures that there are not multiple shortest paths between pairs of vertices. We approximate the path that data takes from a monitor to a target node by the shortest path. This follows the experimental design of [12].

For each graph distribution, and for a range of graph sizes, edge densities, monitor set sizes, and target set sizes, we estimate the degree of every vertex by $\widehat{\deg}(u)$ and by the biased estimator given by the union of the edges discovered by traceroute sampling,

$$\widehat{\deg}_{\text{biased}}(u) = \left| \bigcup_{s \in V_m} N_s(u) \right|,$$

where V_m is the set of monitor nodes and $N_s(u)$ denotes the neighbors of u in the union of all routes discovered when sending packets from s to every node in the target set V_t. This provides estimates of the degree distribution cdf, by the reduced bias estimator $\widehat{d_{\leq k}}$ from above and by the biased estimator $\widehat{d_{\leq k}}_{\text{biased}}$, defined by

$$\widehat{d_{\leq k}}_{\text{biased}} = \frac{\#\{u : \widehat{\deg}_{\text{biased}}(u) \leq k\}}{\#\{u : \widehat{\deg}_{\text{biased}}(u) \geq 1\}}.$$

$\widehat{d_{\leq k}}_{\text{biased}}$ has been the primary approach considered in prior work.

We use these estimates to calculate the ℓ_2 error of the degree distribution cdf estimate, given by

$$\text{err}_{\text{biased}} = \frac{\left(\sum_{k=0}^{\infty} \left(\widehat{d_{\leq k}}_{\text{biased}} - \Pr[\deg(u) \leq k] \right)^2 \right)^{1/2}}{\left(\sum_{k=0}^{\infty} \Pr[\deg(u) \leq k]^2 \right)^{1/2}},$$

and

$$\text{err}_{\text{reduced}} = \frac{\left(\sum_{k=0}^{\infty} \left(\widehat{d_{\leq k}} - \Pr[\deg(u) \leq k] \right)^2 \right)^{1/2}}{\left(\sum_{k=0}^{\infty} \Pr[\deg(u) \leq k]^2 \right)^{1/2}},$$

where $\Pr[\deg(u) \leq k] = \#\{u : \deg(u) \leq k\}/n$ is the probability with respect to a uniformly random choice of u from the vertices of G.

We also exhibit plots of the distribution and the two estimates for a typical parameter setting. All error values reported are the median value of 100 experiments, and the plots show the distribution with the median error as well as the pointwise 90th percentile values from the 100 experiments.

4.1   Random Graph, $G_{n,m}$

The Erdős-Rényi distribution of graphs, $G_{n,m}$, can be generated by choosing a graph uniformly at random from all graphs with n vertices and m edges [6]. It was not developed to model real-world graphs, but it is analytically tractable and can provide insight into the behavior of more realistic graph models. It can also be used as a null hypothesis. Section 3 proved that $\widehat{\deg}(u)$ and $\widehat{d_{\leq k}}$ are asymptotically unbiased for $G_{n,p}$ when $np \gg \log n$. Conventional wisdom holds that anything true for $G_{n,p}$ is also true for $G_{n,m}$ when $m \approx \binom{n}{2}p$, and computer experiments support this conclusion, even for moderately size n and m, as shown in Table 1 and Figure 2a. These experiments indicate that $\widehat{\deg}(u)$ and $\widehat{d_{\leq k}}$ are also good estimators when the number of targets n_t is a reasonably small fraction of n, which is the case in traceroute sampling of the AS graph.

Table 1. ℓ_2 error in degree distribution estimation with and without bias reduction for Erdős-Rényi graph, $G_{n,m}$ where $d = 2m/n$, with n_m monitors and n_t targets (median values of 100 trials)

n	d	n_m	n_t	% err$_{\text{biased}}$	% err$_{\text{reduced}}$
1,000	15	2	$n/8$	3.38	3.15
			$n/2$	3.08	0.96
			n	2.81	0.42
		8	$n/2$	2.11	0.81
		16	$n/2$	1.38	0.80
10,000	20	2	$n/8$	4.02	2.10
			$n/2$	3.75	1.25
			n	3.51	0.46
100,000	15	2	n	2.81	0.21

4.2 Preferential Attachment Graph

The preferential attachment (PA) graph was proposed for a model of the internet and the world wide web by Barabási and Albert in [2], and this has generated a large body of subsequent research, although the validity of the model as a representation of the router graph or the AS graph has been questioned (see, for example, [3]). The estimator $\widehat{\delta_{\leq k}}$ does not perform particularly well on the PA graphs that we used in our experiments, generating ℓ_2 error that is sometimes smaller and sometimes larger than the biased estimator (see Table 2).

The most interesting detail of this series of experiments is the shape of the degree distribution estimated by $\widehat{\delta_{\leq k}}$. When plotted on a log-log scale (Figure 2b),

Table 2. ℓ_2 error in degree distribution estimation with and without bias reduction for Preferential Attachment graph with n nodes and m out-edges per node, n_m monitors and n_t targets (median values of 100 trials)

n	m	n_m	n_t	% err$_{\text{biased}}$	% err$_{\text{reduced}}$
1,000	5	2	$n/8$	2.29	2.35
			$n/2$	1.95	2.66
			n	1.71	2.88
		8	$n/2$	1.26	2.00
		16	$n/2$	0.91	1.57
10,000	10	2	$n/8$	3.47	2.36
			$n/2$	3.23	3.39
			n	3.03	4.31
100,000	15	2	n	3.99	4.43

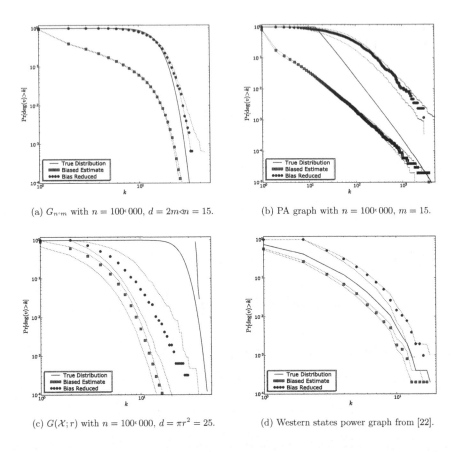

(a) $G_{n,m}$ with $n = 100,000$, $d = 2m/n = 15$.

(b) PA graph with $n = 100,000$, $m = 15$.

(c) $G(\mathcal{X}; r)$ with $n = 100,000$, $d = \pi r^2 = 25$.

(d) Western states power graph from [22].

Fig. 2. Degree sequence ccdf, biased, and bias reduced estimators for synthetic data, with 2 monitor nodes chosen uniformly at random, n target nodes, and shortest path sampling used to approximate traceroute. Plots based on 100 trials, where data points correspond to trial with median ℓ_2 error, and dotted region shows pointwise bounds on 90% of trials.

the biased estimate of the degree distribution appears to be straight line, although with a different slope than the underlying distribution (this is consistent with the theoretical results of [1]). However, the "biased reduced" estimate appears to fall off faster than linear (when plotted on a log-log scale). This is typical of the experiments we conducted with other parameter settings for the PA graph. It could be an effect of the instance sizes being too small, but it persists over two orders of magnitude. Thus, it seems that locally non-tree-like aspects of the PA graph are decreasing the accuracy of $\widehat{\delta_{\leq k}}$. As shown in Figure 1 and to be elaborated upon in Section 5, the degree distribution of the AS graph *does not* fall off faster than linear when estimated with $\widehat{\delta_{\leq k}}$. This could mean that the shortest path routing used in the experiment is not a close enough approximation of the true traceroute sampled paths. But it *could* be interpreted as additional evidence that the AS graph is not distributed according to the PA graph process.

4.3   Random Geometric Graph, $G(\mathcal{X}; r)$

For graphs with high clustering coefficient, the proof sketched in Section 3 will not apply. However, the traceroute paths found by skitter exhibit some level of clustering. To investigate the performance of the bias-reduction technique on graphs with clustering, we examine random geometric graphs $G(\mathcal{X}; r)$. These graphs are formed by selecting a set of n points independently and uniformly at random from the unit square, and linking two points with an edge if and only if they are within ℓ_2 distance r (for a detailed treatment, see [16]). The performance of the bias-reduction technique is summarized for a variety of geometric random graphs in Table 3.

The plot exhibited in Figure 2c is typical for the performance of bias reduction on random geometric graphs; although the bias-reduced estimate is closer to the truth, it is still quite far away from it. The tail of the estimated ccdf, with or without bias reduction, falls off noticeably more slowly than that of the true degree distribution, and looks more like a power-law than it should.

In light of this, it seems that future research should investigate the amount of clustering present in the AS graph. This will permit us to better gauge the accuracy of the bias-reduced estimate of the degree distribution there. However, understanding clustering in the AS graph is hard for the same reasons that understanding the degree distribution is hard, which is due to the lack of unbiased data.

Table 3. ℓ_2 error in degree distribution estimation with and without bias reduction for geometric random graph, $G(\mathcal{X}, r)$ where $d = \pi r^2 n$, with n_m monitors and n_t targets (median values of 100 trials)

n	d	n_m	n_t	$\text{err}_{\text{biased}}$	$\text{err}_{\text{reduced}}$
1,000	15	2	$n/8$	3.14	2.77
			$n/2$	2.91	2.49
			n	2.73	2.17
		8	$n/2$	2.45	2.50
		16	$n/2$	2.23	2.49
10,000	20	2	$n/8$	3.87	3.57
			$n/2$	3.68	3.36
			n	3.55	3.16
100,000	25	2	n	4.19	3.90

4.4   Western States Power Graph

In addition to studying the behavior of bias reduction on the random graphs describe above, we also consider the estimator's performance on synthetic data that is based on a network from the real world, the Western States Power Grid

Table 4. ℓ_2 error in degree distribution estimation with and without bias reduction for Western States Power Graph ($n = 4,941$, $m = 6,594$) with n_m monitors and n_t targets (median values of 100 trials)

$\Pr[\deg(u) \le k]$:

n	d	n_m	n_t	$\mathrm{err}_{\mathrm{biased}}$	$\mathrm{err}_{\mathrm{reduced}}$
4,941	2.67	2	n	0.25	0.75

$\Pr\left[\deg(u) \le k \mid \deg(u) \ge 3\right]$:

n	d	n_m	n_t	$\mathrm{err}_{\mathrm{biased}}$	$\mathrm{err}_{\mathrm{reduced}}$
4,941	2.67	2	$n/8$	0.24	0.13
			$n/2$	0.12	0.06
			n	0.06	0.05
		8	$n/2$	0.09	0.06
		16	$n/2$	0.08	0.09

Graph (WSPG Graph) [22]. This graph represents the power transmission links between 4,941 nodes, representing the generators, transformers, and substations in the Western United States. It is roughly similar in size to the AS graph, and also similar because both networks represent real objects which are connected by real wires.

The result of the bias-reduction technique is shown in Figure 2d. The ℓ_2 error is higher after bias reduction, but this is because the bias-reduction technique filters out all vertices of degree less than 3. Since these low degree vertices are prevalent in the WSPG graph, we also compare the bias-reduced estimate to the degree distribution of the WSPG graph restricted to vertices of degree 3 and higher. Table 4 shows the unconditioned ℓ_2 error for one experiment, and the ℓ_2 error of the estimated cdfs conditioned on vertices having degree at least 3 for a range of experiments.

5   AS Graph

The previous two sections showed theoretically and by computer simulations that the bias-reduction technique developed in Section 2 can be an effective way to reduce the errors introduced by traceroute sampling. This section reports on the results of applying the bias-reduction technique to traceroute-sampled data from the CAIDA skitter project.

A recent paper by Mahadevan, Krioukov, Fomenkov, Dimitropoulos, claffy, and Vahdat provides a detailed analysis of CAIDA skitter data from March, 2004 [14]. We follow the methodology used there, and, in particular, we aggregate the routes observed over the course of a month (from daily graphs provided by CAIDA), and we remove all AS-sets, multi-origin ASes, and private ASes, and discard all indirect links.

The results of applying the bias-reduction technique to the March, 2004 skitter data are plotted in Figure 1. This data set contains 9,204 nodes and 28,959

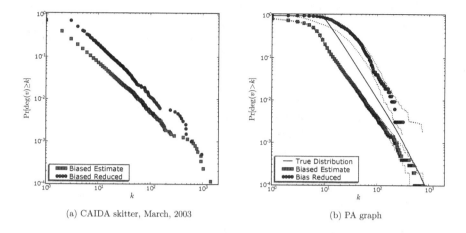

(a) CAIDA skitter, March, 2003   (b) PA graph

Fig. 3. Estimated degree distribution ccdf of CAIDA skitter data from March, 2003 with and without bias reduction and estimated degree distribution ccdf of PA Graph with similar parameters ($n = 10,000$ nodes, $m = 10$ out-edges per node, 20 source nodes, $n/2$ target nodes) with and without bias reduction. Both estimates of skitter data follow power law, but bias reduced estimate of PA Graph does not.

edges, so the average degree before bias reduction is 6.29. There are 22 ASes in the monitor set, and between 10% and 50% of ASes are represented in the target set. The bias-reduction technique yields an estimate of $\widehat{\deg}(u) < \infty$ for $2,078$ vertices, and the average degree after bias reduction is 12.66 (which is very close to 12.52, the biased average degree of vertices with degree at least 3).

The behavior of the bias reduced estimate for k values around 64 is particularly interesting (see Figure 1). Although it is far from definitive, the lack of ASes with degree between 65 and 90 could be the result of economic or technological factors. For example, the Juniper T320 edge router has the ability to house up to 64 interfaces in one chassis. This, or similar product specifications, could lead AS operators to avoid connecting to *slightly* more than 64 other ASes.

Finally, the fact that the bias reduced estimate does *not* fall off at a superlinear rate provides some additional evidence against the theory that the AS graph is an example of a preferential attachment model (see comparison in Figure 3). This argument has been made previously based on completely different considerations (see, for example, [3]).

6   Conclusion

In this paper we introduced a new approach to addressing the bias inherent to traceroute sampling. Starting from the multiple-recapture population estimation technique of statistics, we developed a bias reduction technique applicable to the highly dependent random variables present in path sampling.

In an idealized theoretical framework of shortest path sampling in Erdős-Rényi graphs, we described how to rigorously prove that the proposed estimator is asymptotically unbiased, and, using computer experiments, we show that the estimator can give significant improvements when the target nodes constitute a fraction of vertex set. Computer experiments also highlighted some of the weak points of this estimator, including the less-than-perfect estimates on locally non-tree-like graphs, like the PA graph and the random geometric graph.

Applying the bias-reduction technique to the CAIDA skitter data provided new evidence that the AS graph is not a preferential attachment graph, and also uncovered a way that economic and technological limitations are reflected in the AS degree distribution.

The theoretical and computer simulations supporting the effectiveness of the bias-reduction technique all rely on the assumption that shortest path routing is a close-enough approximation of BGP routing. This assumption should be considered in more detail, and the behavior of the bias-reduction technique under a more realistic model of traceroute is an important future direction of research.

Acknowledgments

ADF would like to thank Josh Grubman for pointing us towards the specifications of the Juniper T320 router, even if he does not believe that product specifications are likely to result in the absence of nodes with degree slightly above 64.

References

1. Achlioptas, D., Clauset, A., Kempe, D., Moore, C.: On the bias of traceroute sampling: or, power-law degree distributions in regular graphs. In: STOC 2005. Proceedings of the thirty-seventh annual ACM symposium on Theory of computing, pp. 694–703. ACM Press, New York, NY, USA (2005)
2. Barabási, A.-L., Albert, R.: Emergence of scaling in random networks. Science 286(5439), 509–512 (1999)
3. Chen, Q., Chang, H., Govindan, R., Jamin, S., Shenker, S.J., Willinger, W.: The origin of power laws in Internet topologies revisited. In: INFOCOM 2002. Proceedings Twenty-First Annual Joint Conference of the IEEE Computer and Communications Societies, vol. 2, pp. 608–617 (2002)
4. Clauset, A., Moore, C.: Accuracy and scaling phenomena in internet mapping. Physical Review Letters 94(1), 18701 (2005)
5. Dall'Asta, L., Alvarez-Hamelin, I., Barrat, A., Vazquez, A., Vespignani, A.: A statistical approach to the traceroute-like exploration of networks: theory and simulations. In: López-Ortiz, A., Hamel, A.M. (eds.) CAAN 2004. LNCS, vol. 3405, Springer, Heidelberg (2005)
6. Erdős, P., Rényi, A.: On random graphs. I. Publ. Math. Debrecen 6, 290–297 (1959)

7. Faloutsos, M., Faloutsos, P., Faloutsos, C.: On power-law relationships of the internet topology. In: SIGCOMM 1999. Proceedings of the conference on Applications, technologies, architectures, and protocols for computer communication, NY, USA, pp. 251–262. ACM Press, New York (1999)
8. Frank, O.: A survey of statistical methods for graph analysis. Sociological Methodology 12, 110–155 (1981)
9. Guillaume, J.-L., Latapy, M., Magoni, D.: Relevance of massively distributed explorations of the internet topology: Qualitative results. Computer networks 50(16), 3197–3224 (2006)
10. claffy, k.c., Monk, T.E., McRobb, D.: Internet tomography. Nature (January 1999)
11. Klovdahl, A.S.: The Small World (in honor of Stanley Milgram), chapter Urban social networks. In: Some methodological problems and possibilities, ABLEX, Greenwich (1989)
12. Lakhina, A., Byers, J., Crovella, M., Xie, P.: Sampling biases in ip topology measurements. In: INFOCOM 2003. 22nd Annual Joint Conference of the IEEE Computer and Communications Societies, vol. 1, pp. 332–341. IEEE, Los Alamitos (2003)
13. Leguay, J., Latapy, M., Friedman, T., Salamatian, K.: Describing and simulating internet routes. In: Boutaba, R., Almeroth, K.C., Puigjaner, R., Shen, S., Black, J.P. (eds.) NETWORKING 2005. LNCS, vol. 3462, pp. 659–670. Springer, Heidelberg (2005)
14. Mahadevan, P., Krioukov, D., Fomenkov, M., Dimitropoulos, X., claffy, k.c., Vahdat, A.: The internet AS-level topology: three data sources and one definitive metric. SIGCOMM Comput. Commun. Rev. 36(1), 17–26 (2006)
15. Pansiot, J.-J., Grad, D.: On routes and multicast trees in the internet. SIGCOMM Comput. Commun. Rev. 28(1), 41–50 (1998)
16. Penrose, M.: Random geometric graphs. In: Oxford Studies in Probability, vol. 5, Oxford University Press, Oxford (2003)
17. Petermann, T., Rios, P.D.L.: Exploration of scale-free networks. European Physical Journal B 38, 201–204 (2004)
18. Pickands, J.I., Raghavachari, M.: Exact and asymptotic inference for the size of a population. Biometrika 74(2), 355–363 (1987)
19. Pólya, G.: Probabilities in proofreading. Amer. Math. Monthly 83(1), 42 (1975)
20. Salganik, M.J., Heckathorn, D.D.: Sampling and estimation in hidden populations using respondent-drive sampling. Sociological Methodology 34, 193–239 (2004)
21. Viger, F., Barrat, A., Dall'Asta, L., Zhang, C., Kolaczyk, E.: Network Inference from TraceRoute Measurements: Internet Topology 'Species'. Phys. Rev. E 75(056111) (2007)
22. Watts, D.J., Strogatz, S.H.: Collective dynamics of "small-world" networks. Nature 292, 440–442 (1998)

Distribution of PageRank Mass Among Principle Components of the Web

Konstantin Avrachenkov, Nelly Litvak, and Kim Son Pham

[1] INRIA Sophia Antipolis, 2004, Route des Lucioles, 06902, France
`k.avrachenkov@sophia.inria.fr`
[2] University of Twente, Dept. of Applied Mathematics, P.O. Box 217,
7500AE Enschede, The Netherlands
`n.litvak@ewi.utwente.nl`
[3] St.Petersburg State University, 35, University Prospect, 198504, Peterhof, Russia
`sonsecure@yahoo.com.sg`

Abstract. We study the PageRank mass of principal components in a bow-tie Web Graph, as a function of the damping factor c. Using a singular perturbation approach, we show that the PageRank share of IN and SCC components remains high even for very large values of the damping factor, in spite of the fact that it drops to zero when $c \to 1$. However, a detailed study of the OUT component reveals the presence of "dead-ends" (small groups of pages linking only to each other) that receive an unfairly high ranking when c is close to one. We argue that this problem can be mitigated by choosing c as small as $1/2$.

1 Introduction

The link-based ranking schemes such as PageRank [1], HITS [2], and SALSA [3] have been successfully used in search engines to provide adequate importance measures for Web pages. In the present work we restrict ourselves to the analysis of the PageRank criterion and use the following definition of PageRank from [4]. Denote by n the total number of pages on the Web and define the $n \times n$ hyper-link matrix W as follows:

$$w_{ij} = \begin{cases} 1/d_i, & \text{if page } i \text{ links to } j, \\ 1/n, & \text{if page } i \text{ is dangling}, \\ 0, & \text{otherwise}, \end{cases} \tag{1}$$

for $i, j = 1, ..., n$, where d_i is the number of outgoing links from page i. A page is called *dangling* if it does not have outgoing links. The PageRank is defined as a stationary distribution of a Markov chain whose state space is the set of all Web pages, and the transition matrix is

$$G = cW + (1-c)(1/n)\mathbf{1}\mathbf{1}^T. \tag{2}$$

Here and throughout the paper we use the symbol $\mathbf{1}$ for a column vector of ones having by default an appropriate dimension. In (2), $\mathbf{1}\mathbf{1}^T$ is a matrix whose all

A. Bonato and F.R.K. Chung (Eds.): WAW 2007, LNCS 4863, pp. 16–28, 2007.

entries are equal to one, and $c \in (0,1)$ is the parameter known as a *damping factor*. Let π be the PageRank vector. Then by definition, $\pi G = \pi$, and $||\pi|| = \pi \mathbf{1} = 1$, where we write $||\mathbf{x}||$ for the L_1-norm of vector $\mathbf{x}$.

The damping factor c is a crucial parameter in the PageRank definition. It regulates the level of the uniform noise introduced to the system. Based on the publicly available information Google originally used $c = 0.85$, which appears to be a reasonable compromise between the true reflection of the Web structure and numerical efficiency (see [5] for more details). However, it was mentioned in [6] that the value of c too close to one results into distorted ranking of important pages. This phenomenon was also independently observed in [7]. Moreover, with smaller c, the PageRank is more robust, that is, one can bound the influence of outgoing links of a page (or a small group of pages) on the PageRank of other groups [8] and on its own PageRank [7].

In this paper we explore the idea of relating the choice of c to specific properties of the Web structure. In papers [9,10] the authors have shown that the Web graph can be divided into three principle components. The Giant Strongly Connected Component (SCC) contains a large group of pages all having a hyperlink path to each other. The pages in the IN (OUT) component have a path to (from) the SCC, but not back. Furthermore, the SCC component is larger than the second largest strongly connected component by several orders of magnitude.

In Section 3 we consider a Markov walk governed by the hyperlink matrix W and explicitly describe the limiting behavior of the PageRank vector as $c \to 1$. We experimentally study the OUT component in more detail to discover a so-called Pure OUT component (the OUT component without dangling nodes and their predecessors) and show that Pure OUT contains a number of small sub-SCC's, or dead-ends, that absorb the total PageRank mass when $c = 1$. In Section 4 we apply the singular perturbation theory [11,12,13,14] to analyze the shape of the PageRank of IN+SCC as a function of c. The dangling nodes turn out to play an unexpectedly important role in the qualitative behavior of this function. Our analytical and experimental results suggest that the PageRank mass of IN+SCC is sustained on a high level for quite large values of c, in spite of the fact that it drops to zero as $c \to 1$. Further, in Section 5 we show that the total PageRank mass of Pure OUT component increases with c. We argue that $c = 0.85$ results in an inadequately high ranking for Pure OUT pages and we present an argument for choosing c as small as $1/2$. We confirm our theoretical argument by experiments with log files. We would like to mention that the value $c = 1/2$ was also used in [15] to find gems in scientific citations. This choice was justified intuitively by stating that researchers may check references in cited papers but on average they hardly go deeper than two levels. Nowadays, when search engines work really fast, this argument also applies to Web search. Indeed, it is easier for the user to refine a query and receive a proper page in fraction of seconds than to look for this page by clicking on hyper-links. Therefore, we may assume that a surfer searching for a page, on average, does not go deeper than two clicks.

The body of the paper contains main ideas and results. An extended version with the necessary information from the perturbation theory and the proofs can be found in [16].

2   Datasets

We have collected two Web graphs, which we denote by INRIA and FrMathInfo. The Web graph INRIA was taken from the site of INRIA, the French Research Institute of Informatics and Automatics. The seed for the INRIA collection was Web page www.inria.fr. It is a typical large Web site with around 300.000 pages and 2 millions hyper-links. We have collected all pages belonging to INRIA. The Web graph FrMathInfo was crawled with the initial seeds of 50 mathematics and informatics laboratories of France, taken from Google Directory. The crawl was executed by Breadth First Search of depth 6. The FrMathInfo Web graph contains around 700.000 pages and 8 millions hyper-links. As the Web seems to have a fractal structure [17], we expect our datasets to be enough representative.

The link structure of the two Web graphs is stored in Oracle database. We could store the adjacency lists in RAM to speed up the computation of PageRank and other quantities of interest. This enables us to make more iterations, which is extremely important when the damping factor c is close to one. Our PageRank computation program consumes about one hour to make 500 iterations for the FrMathInfo dataset and about half an hour for the INRIA dataset for the same number of iterations. Our algorithms for discovering the structures of the Web graph are based on Breadth First Search and Depth First Search methods, which are linear in the sum of number of nodes and links.

3   The Structure of the Hyper-link Transition Matrix

With the bow-tie Web structure [9,10] in mind, we would like to analyze a stationary distribution of a Markov random walk governed by the hyper-link transition matrix W given by (1). Such random walk follows an outgoing link chosen uniformly at random, and dangling nodes are assumed to have links to all pages in the Web. We note that the methods presented below can be easily extended to the case of personalized PageRank [18], when after a visit to a dangling node, the next page is sampled from some prescribed distribution.

Obviously, the graph induced by W has a much higher connectivity than the original Web graph. In particular, if the random walk can move from a dangling node to an arbitrary node with the uniform distribution, then the Giant SCC component increases further in size. We refer to this new strongly connected component as the Extended Strongly Connected Component (ESCC). Due to the artificial links from the dangling nodes, the SCC component and IN component are now inter-connected and are parts of the ESCC. Furthermore, if there are dangling nodes in the OUT component, then these nodes together with all their predecessors become a part of the ESCC.

In the mini-example in Figure 1, node 0 represents the IN component, nodes from 1 to 3 form the SCC component, and the rest of the nodes, nodes from 4 to 11, are in the OUT component. Node 5 is a dangling node, thus, artificial links go from the dangling node 5 to all other nodes. After addition of the artificial links, all nodes from 0 to 5 form the ESCC.

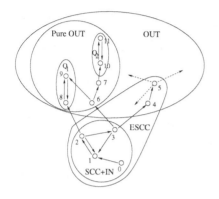

Fig. 1. Example of a graph

#	$INRIA$	$FrMathInfo$
total nodes	318585	764119
nodes in SCC	154142	333175
nodes in IN	0	0
nodes in OUT	164443	430944
nodes in ESCC	300682	760016
nodes in Pure OUT	17903	4103
SCCs in OUT	1148	1382
SCCs in Pure Out	631	379

Fig. 2. Component sizes in INRIA and Fr-MathInfo datasets

In the Markov chain induced by the matrix W, all states from ESCC are *transient*, that is, with probability 1, the Markov chain eventually leaves this set of states and never returns back. The stationary probability of all these states is zero. The part of the OUT component without dangling nodes and their predecessors forms a block that we refer to as a Pure OUT component. In Figure 1 the Pure OUT component consists of nodes from 6 to 11. Typically, the Pure OUT component is much smaller than the Extended SCC. However, this is the set where the total stationary probability mass is concentrated. The sizes of all components for our two datasets are given in Figure 2. Here the size of the IN components is zero because in the Web crawl we used the Breadth First Search method and we started from important pages in the Giant SCC. For the purposes of the present research it does not make any difference since we always consider IN and SCC together.

Let us now analyze the structure of the Pure OUT component in more detail. It turns out that inside Pure OUT there are many disjoint strongly connected components. All states in these sub-SCC's (or, "dead-ends") are *recurrent*, that is, the Markov chain started from any of these states always returns back to it. In particular, we have observed that there are many dead-ends of size 2 and 3. The Pure OUT component also contains transient states that eventually bring the random walk into one of the dead-ends. For simplicity, we add these states to the giant transient ESCC component.

Now, by appropriate renumbering of the states, we can refine the matrix W by subdividing all states into one giant transient block and a number of small recurrent blocks as follows:

$$W = \begin{bmatrix} Q_1 & & 0 & 0 \\ & \ddots & & \\ 0 & & Q_m & 0 \\ R_1 & \cdots & R_m & T \end{bmatrix} \begin{matrix} \text{dead-end (recurrent)} \\ \cdots \\ \text{dead-end (recurrent)} \\ \text{ESCC+[transient states in Pure OUT] (transient)} \end{matrix} \qquad (3)$$

Here for $i = 1, \ldots, m$, a block Q_i corresponds to transitions inside the i-th recurrent block, and a block R_i contains transition probabilities from transient states to the i-th recurrent block. Block T corresponds to transitions between the transient states. For instance, in example of the graph from Figure 1, the nodes 8 and 9 correspond to block Q_1, nodes 10 and 11 correspond to block Q_2, and all other nodes belong to block T. Let us denote by $\bar{\pi}_{OUT,i}$ the stationary distribution corresponding to block Q_i.

We would like to emphasis that the recurrent blocks here are really small, constituting altogether about 5% for INRIA and about 0.5% for FrMathInfo. We believe that for larger data sets, this percentage will be even less. By far most important part of the pages is contained in the ESCC, which constitutes the major part of the giant transient block.

Next, we note that if $c < 1$, then all states in the Markov chain induced by the Google transition matrix (2) are recurrent, which automatically implies that they all have positive stationary probabilities. However, if $c = 1$, the majority of pages turn into transient states with stationary probability zero. Hence, the random walk governed by the Google transition matrix G is in fact a singularly perturbed Markov chain. Informally, by singular perturbation we mean relatively small changes in elements of the matrix, that lead to altered connectivity and stationary behavior of the chain. Using the results of the singular perturbation theory (see e.g., [11,12,13,14]), in the next proposition we characterize explicitly the limiting PageRank vector as $c \to 1$.

Proposition 1. *Let $\bar{\pi}_{OUT,i}$ be a stationary distribution of the Markov chain governed by Q_i, $i = 1, \ldots, m$. Then, we have*

$$\lim_{c \to 1} \pi(c) = [\pi_{OUT,1} \ \cdots \ \pi_{OUT,m} \ \mathbf{0}],$$

where

$$\pi_{OUT,i} = \left(\frac{\# \ nodes \ in \ block \ Q_i}{n} + \frac{1}{n} \mathbf{1}^T [I - T]^{-1} R_i \mathbf{1} \right) \bar{\pi}_{OUT,i} \qquad (4)$$

for $i = 1, ..., m$, I is the identity matrix, and $\mathbf{0}$ is a row vector of zeros that correspond to stationary probabilities of the states in the transient block.

The second term inside the brackets in formula (4) corresponds to the PageRank mass received by a dead-end from the Extended SCC. If c is close to one, then this contribution can outweight by far the fair share of the PageRank, whereas the PageRank mass of the giant transient block decreases to zero. How large is the neighborhood of one where the ranking is skewed towards the Pure OUT?

Is the value $c = 0.85$ already too large? We will address these questions in the remainder of the paper. In the next section we analyze the PageRank mass IN+SCC component, which is an important part of the transient block.

4   PageRank Mass of IN+SCC

In Figure 3 we depict the PageRank mass of the giant component IN+SCC, as a function of the damping factor, for FrMathInfo. Here we see a typical behavior also observed for several pages in the mini-web from [6]: the PageRank first grows with c and then decreases to zero. In our case, the PageRank mass of IN+SCC drops drastically starting from some value c close to one. We can explain this phenomenon by highlighting the role of the dangling nodes.

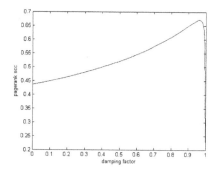

Fig. 3. The PageRank mass of IN+SCC as a function of c

We start the analysis by subdividing the Web graph sample into three subsets of nodes: IN+SCC, OUT, and the set of dangling nodes DN. We assume that no dangling node originates from OUT. This simplifies the derivation but does not change our conclusions. Then the Web hyper-link matrix W in (1) can be written in the form

$$W = \begin{bmatrix} Q & 0 & 0 \\ R & P & S \\ \frac{1}{n}\mathbf{1}\mathbf{1}^T & \frac{1}{n}\mathbf{1}\mathbf{1}^T & \frac{1}{n}\mathbf{1}\mathbf{1}^T \end{bmatrix} \begin{matrix} \text{OUT} \\ \text{IN+SCC} \\ \text{DN} \end{matrix} \;,$$

where the block Q corresponds to the hyper-links inside the OUT component, the block R corresponds to the hyper-links from IN+SCC to OUT, the block P corresponds to the hyper-links inside the IN+SCC component, and the block S corresponds to the hyper-links from SCC to dangling nodes. In the above, n is the total number of pages in the Web graph sample, and the blocks $\mathbf{1}\mathbf{1}^T$ are the matrices of ones adjusted to appropriate dimensions.

Dividing the PageRank vector in segments corresponding to the blocks OUT, IN+SCC and DN, namely, $\pi = [\pi_{\text{OUT}} \; \pi_{\text{IN+SCC}} \; \pi_{\text{DN}}]$, we can rewrite the well-known formula (see e.g. [19])

$$\pi = \frac{1-c}{n}\mathbf{1}^T[I - cW]^{-1} \tag{5}$$

as a system of three linear equations:

$$\pi_{\text{OUT}}[I - cQ] - \pi_{\text{IN+SCC}}cR - \frac{c}{n}\pi_{\text{DN}}\mathbf{1}\mathbf{1}^T = \frac{1-c}{n}\mathbf{1}^T, \tag{6}$$

$$\pi_{\text{IN+SCC}}[I - cP] - \frac{c}{n}\pi_{\text{DN}}\mathbf{1}\mathbf{1}^T = \frac{1-c}{n}\mathbf{1}^T, \tag{7}$$

$$-\pi_{\text{IN+SCC}}cS + \pi_{\text{DN}} - \frac{c}{n}\pi_{\text{DN}}\mathbf{1}\mathbf{1}^T = \frac{1-c}{n}\mathbf{1}^T. \tag{8}$$

Solving (6–8) for $\pi_{\text{IN+SCC}}$ we obtain

$$\pi_{\text{IN+SCC}}(c) = \frac{(1-c)\alpha}{1-c\beta}\mathbf{u}_{\text{IN+SCC}}\left[I - cP - \frac{c^2\alpha}{1-c\beta}S\mathbf{1}\mathbf{u}_{\text{IN+SCC}}\right]^{-1}, \tag{9}$$

where $\alpha = |\text{IN} + \text{SCC}|/n$ and $\beta = |\text{DN}|/n$ are the fractions of nodes in IN+SCC and DN, respectively, and $\mathbf{u}_{\text{IN+SCC}} = |\text{IN} + \text{SCC}|^{-1}\mathbf{1}^T$ is a uniform probability row-vector of dimension $|\text{IN} + \text{SCC}|$. Now, define

$$k(c) = \frac{(1-c)\alpha}{1-c\beta}, \quad \text{and} \quad U(c) = P + \frac{c\alpha}{1-c\beta}S\mathbf{1}\mathbf{u}_{\text{IN+SCC}}. \tag{10}$$

Then the derivative of $\pi_{\text{IN+SCC}}(c)$ with respect to c is given by

$$\pi'_{\text{IN+SCC}}(c) = \mathbf{u}_{\text{IN+SCC}}\left\{k'(c)I + k(c)[I - cU(c)]^{-1}(cU(c))'\right\}[I - cU(c)]^{-1}, \tag{11}$$

where using (10) after simple calculations we get $k'(c) = -(1-\beta)\alpha/(1-c\beta)^2$, $(cU(c))' = U(c) + c\alpha/(1-c\beta)^2 S\mathbf{1}\mathbf{u}_{\text{IN+SCC}}$. Let us consider the point $c = 0$. Using (11), we obtain

$$\pi'_{\text{IN+SCC}}(0) = -\alpha(1-\beta)\mathbf{u}_{\text{IN+SCC}} + \alpha\mathbf{u}_{\text{IN+SCC}}P. \tag{12}$$

One can see from the above equation that the PageRank of pages in IN+SCC with many incoming links will increase as c increases from zero, which explains the graphs presented in [6].

Next, let us analyze the total mass of the IN+SCC component. From (12) we obtain $||\pi'_{\text{IN+SCC}}(0)|| = -\alpha(1-\beta)\mathbf{u}_{\text{IN+SCC}}\mathbf{1} + \alpha\mathbf{u}_{\text{IN+SCC}}P\mathbf{1} = \alpha(-1+\beta+p_1)$, where $p_1 = \mathbf{u}_{\text{IN+SCC}}P\mathbf{1}$ is the probability that a random walk stays in IN+SCC for one step if the initial distribution is uniform over IN+SCC. If $1 - \beta < p_1$ then the derivative at 0 is positive. Since dangling nodes typically constitute more than 25% of the graph [20], and p_1 is usually close to one, the condition $1 - \beta < p_1$ seems to be comfortably satisfied in Web samples. Thus, the total PageRank of the IN+SCC increases in c when c is small. Note by the way that if $\beta = 0$ then $||\pi_{\text{IN+SCC}}(c)||$ is strictly decreasing in c. Hence, surprisingly, the presence of dangling nodes qualitatively changes the behavior of the IN+SCC PageRank mass.

Now let us consider the point $c = 1$. Again using (11), we obtain

$$\pi'_{\text{IN+SCC}}(1) = -\frac{\alpha}{1-\beta}\mathbf{u}_{\text{IN+SCC}}[I - P - \frac{\alpha}{1-\beta}S\mathbf{1}\mathbf{u}_{\text{IN+SCC}}]^{-1}. \tag{13}$$

Note that the matrix in the square braces is close to singular. Denote by $\bar{P}$ the hyper-link matrix of IN+SCC when the outer links are neglected. Then, $\bar{P}$ is an irreducible stochastic matrix. Denote its stationary distribution by $\bar{\pi}_{\text{IN+SCC}}$. Then we can apply Lemma A.1 of [16] from the singular perturbation theory to (13) by taking $A = \bar{P}$, $\varepsilon C = \bar{P} - P - \alpha/(1-\beta)S\mathbf{1}\mathbf{u}_{\text{IN+SCC}}$, and noting that $\varepsilon C\mathbf{1} = R\mathbf{1} + (1 - \alpha - \beta)(1 - \beta)^{-1}S\mathbf{1}$. Combining all terms together and using $\bar{\pi}_{\text{IN+SCC}}\mathbf{1} = ||\bar{\pi}_{\text{IN+SCC}}|| = 1$ and $\mathbf{u}_{\text{IN+SCC}}\mathbf{1} = ||\mathbf{u}_{\text{IN+SCC}}|| = 1$, by Lemma A.1 of [16] we obtain

$$||\pi'_{\text{IN+SCC}}(1)|| \approx -\frac{\alpha}{1-\beta}\frac{1}{\bar{\pi}_{\text{IN+SCC}}R\mathbf{1} + \frac{1-\beta-\alpha}{1-\beta}\bar{\pi}_{\text{IN+SCC}}S\mathbf{1}}.$$

It is expected that the value of $\bar{\pi}_{\text{IN+SCC}}R\mathbf{1} + \frac{1-\beta-\alpha}{1-\beta}\bar{\pi}_{\text{IN+SCC}}S\mathbf{1}$ is typically small (indeed, in our dataset INRIA, the value is 0.022), and hence the mass $||\pi_{\text{IN+SCC}}(c)||$ decreases very fast as c approaches one.

Having described the behavior of the PageRank mass $||\pi_{\text{IN+SCC}}(c)||$ at the boundary points $c = 0$ and $c = 1$, now we would like to show that there is at most one extremum on $(0,1)$. It is sufficient to prove that if $||\pi'_{\text{IN+SCC}}(c_0)|| \le 0$ for some $c_0 \in (0,1)$ then $||\pi'_{\text{IN+SCC}}(c)|| \le 0$ for all $c > c_0$. To this end, we apply the Sherman-Morrison formula to (9), which yields

$$\pi_{\text{IN+SCC}}(c) = \tilde{\pi}_{\text{IN+SCC}}(c) + \frac{\frac{c^2\alpha}{1-c\beta}\mathbf{u}_{\text{IN+SCC}}[I - cP]^{-1}S\mathbf{1}}{1 + \frac{c^2\alpha}{1-c\beta}\mathbf{u}_{\text{IN+SCC}}[I - cP]^{-1}S\mathbf{1}}\tilde{\pi}_{\text{IN+SCC}}(c), \tag{14}$$

where

$$\tilde{\pi}_{\text{IN+SCC}}(c) = \frac{(1-c)\alpha}{1-c\beta}\mathbf{u}_{\text{IN+SCC}}[I - cP]^{-1}. \tag{15}$$

represents the main term in the right-hand side of (14). (The second summand in (14) is about 10% of the total sum for the INRIA dataset for $c = 0.85$.) Now the behavior of $\pi_{\text{IN+SCC}}(c)$ in Figure 3 can be explained by means of the next proposition.

Proposition 2. *The term* $||\tilde{\pi}_{\text{IN+SCC}}(c)||$ *given by (15) has exactly one local maximum at some* $c_0 \in [0,1]$. *Moreover,* $||\tilde{\pi}''_{\text{IN+SCC}}(c)|| < 0$ *for* $c \in (c_0, 1]$.

We conclude that $||\tilde{\pi}_{\text{IN+SCC}}(c)||$ is decreasing and concave for $c \in [c_0, 1]$, where $||\tilde{\pi}'_{\text{IN+SCC}}(c_0)|| = 0$. This is exactly the behavior we observe in the experiments. The analysis and experiments suggest that c_0 is definitely larger than 0.85 and actually is quite close to one. Thus, one may want to choose large c in order to maximize the PageRank mass of IN+SCC. However, in the next section we will indicate important drawbacks of this choice.

5   PageRank Mass of ESCC

Let us now consider the PageRank mass of the Extended SCC component (ESCC) described in Section 3, as a function of $c \in [0, 1]$. Subdividing the Page-Rank vector in the blocks $\pi = [\pi_{\text{PureOUT}} \ \pi_{\text{ESCC}}]$, from (5) we obtain

$$||\pi_{\text{ESCC}}(c)|| = (1 - c)\gamma \mathbf{u}_{\text{ESCC}}[I - cT]^{-1}\mathbf{1}, \tag{16}$$

where T represents the transition probabilitites inside the ESCC block, $\gamma = |\text{ESCC}|/n$, and $\mathbf{u}_{\text{ESCC}}$ is a uniform probability row-vector over ESCC. Clearly, we have $||\pi_{\text{ESCC}}(0)|| = \gamma$ and $||\pi_{\text{ESCC}}(1)|| = 0$. Furthermore, by taking derivatives we easily show that $||\pi_{\text{ESCC}}(c)||$ is a concave decreasing function. In the next proposition, we derive upper and lower bounds for $||\pi_{\text{ESCC}}(c)||$.

Proposition 3. *Let λ_1 be the Perron-Frobenius eigenvalue of T, and let $p_1 = \mathbf{u}_{ESCC}T\mathbf{1}$ be the probability that the random walk started from a randomly chosen state in ESCC, stays in ESCC for one step.*

(i) If $p_1 < \lambda_1$ then

$$||\pi_{ESCC}(c)|| < \frac{\gamma(1 - c)}{1 - c\lambda_1}, \quad c \in (0, 1). \tag{17}$$

(ii) If $1/(1 - p_1) < \mathbf{u}_{ESCC}[I - T]^{-1}\mathbf{1}$ then

$$||\pi_{ESCC}(c)|| > \frac{\gamma(1 - c)}{1 - cp_1}, \quad c \in (0, 1). \tag{18}$$

The condition $p_1 < \lambda_1$ has a clear intuitive interpretation. Let $\hat{\pi}_{\text{ESCC}}$ be the probability-normed left Perron-Frobenius eigenvector of T. Then $\hat{\pi}_{\text{ESCC}}$, also known as a *quasi-stationary* distribution of T, is the limiting probability distribution of the Markov chain given that the random walk never leaves the block T (see e.g. [21]). Since $\hat{\pi}_{\text{ESCC}}T = \lambda_1 \hat{\pi}_{\text{ESCC}}$, the condition $p_1 < \lambda_1$ means that the chance to stay in ESCC for one step in the quasi-stationary regime is higher than starting from the uniform distribution $\mathbf{u}_{\text{ESCC}}$. Although $p_1 < \lambda_1$ does not hold in general, one may expect that it should hold for transition matrices describing large entangled graphs since quasi-stationary distribution should favor states, from which the chance to leave ESCC is lower.

Both conditions of Proposition 3 are satisfied in our experiments. With the help of the derived bounds we conclude that $||\pi_{\text{ESCC}}(c)||$ decreases very slowly for small and moderate values of c, and it decreases extremely fast when c becomes close to 1. This typical behavior is clearly seen in Figure 4, where $||\pi_{\text{ESCC}}(c)||$ is plotted with a solid line. The bounds are plotted in Figure 4 with dashed lines. For the INRIA dataset we have $p_1 = 0.97557$, $\lambda_1 = 0.99954$, and for the FrMathInfo dataset we have $p_1 = 0.99659$, $\lambda_1 = 0.99937$.

From the above we conclude that the PageRank mass of ESCC is smaller than γ for any value $c > 0$. On contrary, the PageRank mass of Pure OUT increases in c beyond its "fair share" $\delta = |\text{PureOUT}|/n$. With $c = 0.85$, the PageRank mass of the Pure OUT component in the INRIA dataset is equal to 1.95δ. In

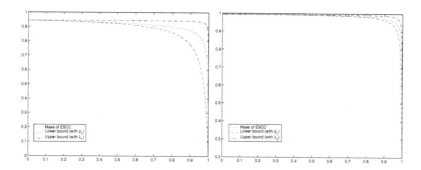

Fig. 4. PageRank mass of ESCC and bounds, INRIA (left) and FrMathInfo (right)

the FrMathInfo dataset, the unfairness is even more pronounced: the PageRank mass of the Pure OUT component is equal to 3.44δ. This gives users an incentive to create dead-ends: groups of pages that link only to each other. Clearly, this can be mitigated by choosing a smaller damping factor. Below we propose one way to determine an "optimal" value of c.

Let $\mathbf{v}$ be some probability vector over ESCC. We would like to choose $c = c^*$ that satisfies the condition

$$||\pi_{\mathrm{ESCC}}(c)|| = ||\mathbf{v}T||, \tag{19}$$

that is, starting from $\mathbf{v}$, the probability mass preserved in ESCC after one step should be equal to the PageRank of ESCC. One can think for instance of the following three reasonable choices of $\mathbf{v}$: 1) $\hat{\pi}_{\mathrm{ESCC}}$, the quasi-stationary distribution of T, 2) the uniform vector $\mathbf{u}_{\mathrm{ESCC}}$, and 3) the normalized PageRank vector $\pi_{\mathrm{ESCC}}(c)/||\pi_{\mathrm{ESCC}}(c)||$. The first choice reflects the proximity of T to a stochastic matrix. The second choice is inspired by definition of PageRank (restart from uniform distribution), and the third choice combines both these features.

If conditions of Proposition 3 are satisfied, then (17) and (18) hold, and thus the value of c^* satisfying (19) must be in the interval (c_1, c_2), where

$$(1 - c_1)/(1 - p_1 c_1) = ||\mathbf{v}T||, \quad (1 - c_2)/(1 - \lambda_1 c_2) = ||\mathbf{v}T||.$$

Numerical results for all three choices of $\mathbf{v}$ are presented in Table 1.

If $\mathbf{v} = \hat{\pi}_{\mathrm{ESCC}}$ then we have $||\mathbf{v}T|| = \lambda_1$, which implies $c_1 = (1 - \lambda_1)/(1 - \lambda_1 p_1)$ and $c_2 = 1/(\lambda_1 + 1)$. In this case, the upper bound c_2 is only slightly larger than $1/2$ and c^* is close to zero in our data sets (see Tabel 1). Such small c however leads to ranking that takes into account only local information about the Web graph (see e.g. [22]). The choice $\mathbf{v} = \hat{\pi}_{\mathrm{ESCC}}$ does not seem to represent the dynamics of the system; probably because the "easily bored surfer" random walk that is used in PageRank computations never follows a quasi-stationary distribution since it often restarts itself from the uniform probability vector.

For the uniform vector $\mathbf{v} = \mathbf{u}_{\mathrm{ESCC}}$, we have $||\mathbf{v}T|| = p_1$, which gives c_1, c_2, c^* presented in Table 1. We have obtained a higher upper bound but the values of c^* are still much smaller than 0.85.

Table 1. Values of c^* with bounds

v	c	INRIA	FrMathInfo
$\hat{\pi}_{\text{ESCC}}$	c_1	0.0184	0.1956
	c_2	0.5001	0.5002
	c^*	.02	.16
$\mathbf{u}_{\text{ESCC}}$	c_1	0.5062	0.5009
	c_2	0.9820	0.8051
	c^*	.604	.535
$\pi_{\text{ESCC}}/\|\pi_{\text{ESCC}}\|$	$1/(1+\lambda_1)$	0.5001	0.5002
	$1/(1+p_1)$	0.5062	0.5009

Finally, for the normalized PageRank vector $\mathbf{v} = \pi_{\text{ESCC}}/\|\pi_{\text{ESCC}}\|$, using (16), we rewrite (19) as

$$\|\pi_{\text{ESCC}}(c)\| = \frac{\gamma}{\|\pi_{\text{ESCC}}(c)\|}\pi_{\text{ESCC}}(c)T\mathbf{1} = \frac{\gamma^2(1-c)}{\|\pi_{\text{ESCC}}(c)\|}\mathbf{u}_{\text{IN+SCC}}[I - cT]^{-1}T\mathbf{1},$$

Multiplying by $\|\pi_{\text{ESCC}}(c)\|$, after some algebra we obtain

$$\|\pi_{\text{ESCC}}(c)\|^2 = \frac{\gamma}{c}\|\pi_{\text{ESCC}}(c)\| - \frac{(1-c)\gamma^2}{c}.$$

Solving the quadratic equation for $\|\pi_{\text{ESCC}}(c)\|$, we get

$$\|\pi_{\text{ESCC}}(c)\| = r(c) = \begin{cases} \gamma & \text{if } c \leq 1/2, \\ \frac{\gamma(1-c)}{c} & \text{if } c > 1/2. \end{cases}$$

Hence, the value c^* solving (19) corresponds to the point where the graphs of $\|\pi_{\text{ESCC}}(c)\|$ and $r(c)$ cross each other. There is only one such point on $(0,1)$, and since $\|\pi_{\text{ESCC}}(c)\|$ decreases very slowly unless c is close to one, whereas $r(c)$ decreases relatively fast for $c > 1/2$, we expect that c^* is only slightly larger than $1/2$. Under conditions of Proposition 3, $r(c)$ first crosses the line $\gamma(1-c)/(1-\lambda_1 c)$, then $\|\pi_{\text{ESCC}}(c)\|_1$, and then $\gamma(1-c)/(1-p_1 c)$. Thus, we yield $(1+\lambda_1)^{-1} < c^* < (1+p_1)^{-1}$. Since both λ_1 and p_1 are large, this suggests that c should be chosen around $1/2$. This is also reflected in Tabel 1.

Last but not least, to support our theoretical argument about the undeserved high ranking of pages from Pure OUT, we carry out the following experiment. In the INRIA dataset we have chosen an absorbing component in Pure OUT consisting just of two nodes. We have added an artificial link from one of these nodes to a node in the Giant SCC and recomputed the PageRank. In Table 2 in the column "PR rank w/o link" we give a ranking of a page according to the PageRank value computed before the addition of the artificial link and in the column "PR rank with link" we give a ranking of a page according to the PageRank value computed after the addition of the artificial link. We have also analyzed the log file of the site INRIA Sophia Antipolis (www-sop.inria.fr) and ranked the pages according to the number of clicks for the period of one year up to May 2007. We note that since we have the access only to the log file of the INRIA Sophia Antipolis site, we use the PageRank ranking also only for the pages from the INRIA Sophia Antipolis site. For instance, for $c = 0.85$, the ranking of Page A without an artificial link is 731 (this means that 730 pages are ranked better than Page A among the pages of INRIA Sophia Antipolis). However, its ranking according to the number of clicks is much lower, 2588. This confirms our conjecture that the nodes in Pure OUT obtain unjustifiably high ranking. Next we note that the addition of an artificial link significantly diminishes the ranking. In fact, it brings it close to the ranking provided by the number of clicks. Finally, we draw the attention of the reader to the fact that choosing $c = 1/2$ also significantly reduces the gap between the ranking by PageRank and the ranking by the number of clicks.

Table 2. Comparison between PR and click based rankings

c	PR rank w/o link	PR rank with link	rank by no. of clicks
Node A			
0.5	1648	2307	2588
0.85	731	2101	2588
0.95	226	2116	2588
Node B			
0.5	1648	4009	3649
0.85	731	3279	3649
0.95	226	3563	3649

To summarize, our results indicate that with $c = 0.85$, the Pure OUT component receives an unfairly large share of the PageRank mass. Remarkably, in order to satisfy any of the three intuitive criteria of fairness presented above, the value of c should be drastically reduced. The experiment with the log files confirms the same. Of course, a drastic reduction of c also considerably accelerates the computation of PageRank by numerical methods [23,5,24].

Acknowledgments

This work is supported by EGIDE ECO-NET grant no. 10191XC and by NWO Meervoud grant no. 632.002.401.

References

1. Page, L., Brin, S., Motwani, R., Winograd, T.: The PageRank citation ranking: Bringing order to the Web. Technical report, Stanford University (1998)
2. Kleinberg, J.M.: Authoritative sources in a hyperlinked environment. Journal of the ACM 46(5), 604–632 (1999)
3. Lempel, R., Moran, S.: The stochastic approach for link-structure analysis (SALSA) and the TKC effect. Comput. Networks 33(1-6), 387–401 (2000)

4. Langville, A.N., Meyer, C.D.: Deeper inside PageRank. Internet Math. 1, 335–380 (2003)
5. Langville, A.N., Meyer, C.D.: Google's PageRank and beyond: the science of search engine rankings. Princeton University Press, Princeton, NJ (2006)
6. Boldi, P., Santini, M., Vigna, S.: PageRank as a function of the damping factor. In: Proc. of the Fourteenth International World Wide Web Conference, Chiba, Japan, ACM Press, New York (2005)
7. Avrachenkov, K., Litvak, N.: The effect of new links on Google PageRank. Stoch. Models 22(2), 319–331 (2006)
8. Bianchini, M., Gori, M., Scarselli, F.: Inside PageRank. ACM Trans. Inter. Tech. 5(1), 92–128 (2005)
9. Broder, A., Kumar, R., Maghoul, F., Raghavan, P., Rajagopalan, S., Statac, R., Tomkins, A., Wiener, J.: Graph structure in the Web. Computer Networks 33, 309–320 (2000)
10. Kumar, R., Raghavan, P., Rajagopalan, S., Sivakumar, D., Tomkins, A., Upfal, E.: The Web as a graph. In: PODS 2000. Proc. 19th ACM SIGACT-SIGMOD-AIGART Symp. Principles of Database Systems, pp. 1–10. ACM Press, New York (2000)
11. Avrachenkov, K.: Analytic Perturbation Theory and its Applications. PhD thesis, University of South Australia (1999)
12. Korolyuk, V.S., Turbin, A.F.: Mathematical foundations of the state lumping of large systems. Mathematics and its Applications, vol. 264. Kluwer Academic Publishers, Dordrecht (1993)
13. Pervozvanskii, A.A., Gaitsgori, V.G.: Theory of Suboptimal Decisions. Mathematics and its Applications (Soviet Series), vol. 12. Kluwer Academic Publishers, Dordrecht (1988)
14. Yin, G.G., Zhang, Q.: Discrete-time Markov chains. Applications of Mathematics (New York), vol. 55. Springer, New York (2005)
15. Chen, P., Xie, H., Maslov, S., Redner, S.: Finding scientific gems with Google. Arxiv preprint Physics 0604130 (2006)
16. Avrachenkov, K., Litvak, N., Pham, K.: Distribution of PageRank mass among principle components of the Web. Arxiv preprint CS 0709.2016 (2007)
17. Dill, S., Kumar, R., McCurley, K.S., Rajagopalan, S., Sivakumar, D., Tomkins, A.: Self-similarity in the Web. ACM Trans. Inter. Tech. 2(3), 205–223 (2002)
18. Haveliwala, T.: Topic-sensitive PageRank: A context-sensitive ranking algorithm for Web search. IEEE Transactions on Knowledge and Data Engineering 15(4), 784–796 (2003)
19. Moler, C., Moler, K.: Numerical Computing with MATLAB. In: SIAM (2003)
20. Eiron, N., McCurley, K., Tomlin, J.: Ranking the Web frontier. In: WWW 2004: Proceedings of the 13th international conference on World Wide Web, pp. 309–318. ACM Press, New York (2004)
21. Seneta, E.: Non-negative Matrices and Markov Chains. Springer Series in Statistics. Springer, New York, Revised reprint of the second (1981) edition [Springer-Verlag, New York MR0719544] (2006)
22. Fortunato, S., Flammini, A.: Random walks on directed networks: the case of PageRank. Arxiv preprint Physics 0604203 (2006)
23. Avrachenkov, K., Litvak, N., Nemirovsky, D., Osipova, N.: Monte Carlo methods in PageRank computation: When one iteration is sufficient. SIAM J. Numer. Anal. 45(2), 890–904 (2007)
24. Berkhin, P.: A survey on PageRank computing. Internet Math. 2, 73–120 (2005)

Finding a Dense-Core in Jellyfish Graphs

Mira Gonen, Dana Ron*, Udi Weinsberg, and Avishai Wool

Tel-Aviv University, Ramat Aviv, Israel
gonenmir@post.tau.ac.il, danar@eng.tau.ac.il,
udiw@eng.tau.ac.il, yash@acm.org

Abstract. The connectivity of the Internet crucially depends on the relationships between thousands of Autonomous Systems (ASes) that exchange routing information using the Border Gateway Protocol (BGP). These relationships can be modeled as a graph, called the AS-graph, in which the vertices model the ASes, and the edges model the peering arrangements between the ASes. Based on topological studies, it is widely believed that the Internet graph contains a central dense-core: Informally, this is a small set of high-degree, tightly interconnected ASes that participate in a large fraction of end-to-end routes. Finding this dense-core is a very important practical task when analyzing the Internet's topology.

In this work we introduce a randomized *sublinear* algorithm that finds a dense-core of the AS-graph. We mathematically prove the correctness of our algorithm, bound the density of the core it returns, and analyze its running time. We also implemented our algorithm and tested it on real AS-graph data. Our results show that the core discovered by our algorithm is nearly identical to the cores found by existing algorithms - at a fraction of the running time.

1 Introduction

1.1 Background and Motivation

The connectivity of the Internet crucially depends on the relationships between thousands of Autonomous Systems (ASes) that exchange routing information using the Border Gateway Protocol (BGP). These relationships can be modeled as a graph, called the AS-graph, in which the vertices model the ASes, and the edges model the peering arrangements between the ASes.

Significant progress has been made in the study of the AS-graph's topology over the last few years. A great deal of effort has been spent measuring topological features of the Internet. Numerous research projects [14, 1, 13, 25, 35, 36, 9, 5, 4, 24, 30, 26, 7, 6, 28, 34, 32, 33, 21, 8, 29, 31] have ventured to capture the Internet's topology. Based on these and other topological studies, it is widely believed that the Internet graph contains a central dense-core: Informally, this is a small set of high-degree, tightly interconnected ASes that participate in a large fraction of end-to-end routes. Finding this dense-core is a very important practical task when analyzing the Internet's topology.

There are several ways to define a dense-core precisely, and various corresponding algorithms and heuristics. In the next subsection we briefly survey known definitions

---

* This work was supported by the Israel Science Foundation (grant number 89/05).

A. Bonato and F.R.K. Chung (Eds.): WAW 2007, LNCS 4863, pp. 29–40, 2007.

and algorithms, and shortly discuss their pros and cons. The goal of our work is to describe an algorithm that finds the dense-core (using a reasonable definition of a dense core), is amenable to rigorous mathematical analysis, and is efficient, both asymptotically and when implemented and tested on real AS data.

1.2 Defining a Dense-Core

An early conceptual model for the Internet topology was suggested by Tauro et al. [34]. This work seems to have coined the "jellyfish" term. The authors argued that the Internet topology resembles a jellyfish where the Internet core corresponds to the middle of the cap, which is surrounded by many "tentacles".

The simplest working definition of a dense-core is from Siganos et al. [31]: according to this work, a core is a clique of maximum size. Since the MaxClique problem is NP-hard and is even hard to approximate [22], the authors suggest a greedy algorithm, which we call GreedyMaxClique: Select the highest degree node as the first member of the core. Then, examine each node in decreasing degree order, and add to the core any node that neighbors *all* the nodes already in the core. This algorithm has a time complexity of $O(|E|)$ (where $|E|$ is the number of edges in the graph). On real AS data (with $n \approx 20,000$ and $|E| \approx 60,000$) the algorithm finds a clique of size 13. Since the graph is sparse (that is, $|E| = O(n)$), the algorithm works quite fast. However, the definition of the core as a clique is very restrictive, since it requires 100% edge density[1], and there is no guarantee that the algorithm will indeed find even an approximately maximum clique. In this work we shall refer to such a clique as the *nucleus* of the AS-graph, to distinguish it from other definitions.

Carmi et al. [10, 11] give a different definition for a dense-core. According to their definition, a k-dense-core is a *maximal* set of nodes with degrees $> k$, where the degree is measured in the subgraph induced by the core nodes. Alvarez-Hamelin et al. [2] use a similar k-core decomposition. Carmi et al. [10, 11] described an algorithm to iteratively compute a k-core, which we refer to as the kCore algorithm. For a given minimal degree k, kCore repeatedly eliminates nodes with (residual) degrees $\leq k$, until no more nodes can be eliminated—and the remaining nodes form a k-core. On real AS-graph data, with $k = 30$, they get a core of about 100 nodes. The kCore algorithm has a theoretical time complexity[2] of $O(n^2)$, and in practice it is significantly slower than the GreedyMaxClique algorithm of [31]. Note that even though the algorithm claims to find a "dense core", it is really based on degrees and has a rather weak guarantee about the density of the resulting core: for a degree k, if the discovered core is C then the edge density is $> k/(|C| - 1)$. Furthermore, *a-priori* there is no guarantee on the size of the core that is found or on the discovered density: for a fixed degree k, one can construct an infinite family of connected graphs in which all nodes have degree $\geq k$ and the core density tends to 0.[3]

---

[1] The density of a subgraph with k vertices is the fraction of the $k(k-1)/2$ possible edges that exist in the subgraph.

[2] A more careful implementation, using a bucket-based priority queue, gives complexity $O(n \log n)$.

[3] E.g., take a collection of m k-cliques and connect them via m additional edges. All nodes have a degree of k or $k+1$ so the core is the whole graph. As m grows the density vanishes.

Subramanian et al. [32] suggested a 5-tier hierarchical layering of the AS-graph. Their dense-core - the top tier, is defined as a subset of ASes whose edge density is > 50%. Their tiering agrees with the jellyfish model of [34] in that they, implicitly, assume a single dense-core. They use a simple greedy algorithm for finding (their definition of) a dense-core. However, they report finding a dense-core of only 20 ASes. A similar approach was suggested by Ge et al. [18].

Feige et al. [15] consider the k-densest subgraph problem, which is defined as follows. Given a graph $G = (V, E)$, find a subgraph $H = (X, F)$ of G such that $|X| = k$ and $|F|$ is maximized. This problem is NP-hard. Feige et al. [15] describe an approximation algorithm that gives a ratio of $O(n^\delta)$ for some $\delta < 1/3$. For any particular value of k the greedy algorithm of Asahiro et al. [3] (which is similar to the kCore algorithm) gives the ratio $O(n/k)$. For some specific values of k there are algorithms that produce approximation ratios that are better than $O(n/k)$ [16, 17]. Charikar [12] considers the related problem of finding a subgraph of maximum average degree. He shows that a simple (linear time) greedy algorithm, which is a variant of the kCore algorithm and the algorithm of Asahiro et al. [3], gives a factor-2 approximation. The proof is based on the relation between the greedy algorithm and the dual of the LP formulation of the problem. We note that in general, a subset of (approximately) maximum average degree might be quite different from the notion we are interested in of a relatively small, very dense subgraph. The example given in Footnote 3 illustrates this.

Sagie and Wool [29] suggested an approach that is based on dense k-subgraphs (DkS). They use parts of the DkS approximation algorithm of [15]. On a sampled AS-graph (based on BGP data) their algorithm found a dense-core of 43 ASes, with density 70%. The time complexity of their algorithm is rather high: $O(n^3)$. Bar et al. [4, 5] use the same approach for finding a dense-core.

Against this backdrop of diverging definitions, our goal was to design an algorithm that (i) is not limited to finding a fully-connected clique, (ii) provides a precise density guarantee for the discovered core, (iii) is very efficient, both asymptotically and in practice, and (iv) is amenable to mathematical analysis.

1.3   Contributions

We chose to use a natural definition of a dense-core that focuses on the actual edge density: We define a dense-core as a set of vertices of size k with a given density α. Motivated by graph property-testing algorithms [19], our approach is to use randomized sampling techniques to find such a core. A related approach was applied in [27] to find large conjunctive clusters. However, intuitively, a dense-core in a general graph is a "local" phenomenon, so random sampling has a very low success probability (e.g., if the core is log-sized). Therefore, we restrict ourselves to the practically-interesting class of Jellyfish graphs: Graphs that contain a dense-core - and this core is also well connected to other parts of the graph.

The extra structure provided by Jellyfish graphs is the basis for our main contribution: a *sublinear* randomized algorithm for finding a dense-core in the AS-graph. We rigorously prove the correctness of our algorithm, and the density of the dense-core produced by the algorithm under mild structural assumptions on the graph (assumptions that do hold for the real AS graph).

We implemented our algorithm (JellyCore) and tested it extensively on AS-graph data collected by the DIMES project [30]. We also implemented the kCore algorithm of Carmi et al. [10], and the GreedyMaxClique algorithm of Siganos et al. [31]. On the AS-graph our JellyCore algorithm finds a 60-80%-dense-core, which has a 90% overlap with the core reported by kCore—but JellyCore runs 6 times faster. Furthermore, we also define a *nucleus* within the dense-core as a subset of the highest degree vertices in the dense-core. The nucleus produced by our algorithm has an 80-90% overlap with the 13-node clique reported by GreedyMaxClique - i.e., we find a nucleus containing around 11 of the 13 members in the clique.

Organization: In Section 2 we give definitions and notations. In Section 3 we describe and analyze a simple randomized algorithm (JellyCore) for finding the dense-core, which serves as a basis for our sublinear algorithm. In Section 4 we modify the JellyCore algorithm to a sublinear algorithm. In Section 5 we give an implementation of the JellyCore algorithm and compare it to the algorithms of Carmi et al. [10] and Siganos et al. [31]. We summarize our conclusions in Section 6.

2 Definitions and Notations

Throughout the paper we consider sparse graphs $G = (V, E)$, i.e., $|E| = O(n)$, where $n = |V|$. For the purpose of time complexity analysis, we assume that for every vertex in the graph we know the degree (in $O(1)$ time). We start by some technical definitions leading up to the definition of the dense-core, and the family of Jellyfish graphs.

Definition 1. Closeness to a clique: *Let C^k denote the k-vertex clique. Denote by $dist(G, C^k)$ the distance (as a fraction of $\binom{k}{2}$) between a graph G over k vertices and C^k. Namely, if $dist(G, C^k) = \epsilon$ then $\epsilon\binom{k}{2}$ edges should be added in order to make G into a clique. A graph G over k vertices is ϵ-close to being a clique if $dist(G, C^k) \leq \epsilon$.*

Definition 2. (k, ϵ)-dense-core: *consider a graph G. A subset of k vertices in the graph is a (k, ϵ)-dense-core if the subgraph induced by this set is ϵ-close to a clique.*

Definition 3. *Let C be a subset of vertices of a graph G. The d-nucleus of C, denoted by H, is the subset of vertices of C with degree (not induced degree) at least d.*

For a set of vertices X, let $\Gamma(X)$ denote the set of vertices that neighbor at least one vertex in X, and let $\Gamma_\delta(X)$ denote the set of vertices that neighbor all but at most $\delta|X|$ vertices in X. We next introduce our main definition.

Definition 4. (k, d, c, ϵ)-Jellyfish subgraph: *For integers k and d, and for $0 \leq \epsilon \leq 1$ (that may all be functions of n), and for a constant $c \geq 1$, a graph G contains a (k, d, c, ϵ)-Jellyfish subgraph if it contains a subset C of vertices, with $|C| = k$, that is a (k, ϵ)-dense-core, which has a non-empty d-nucleus H s.t. the following conditions hold:*

1. *For all $v \in C$, v neighbors at least $(1 - \epsilon)|H|$ vertices in H,*
2. *For all but $\epsilon|\Gamma_{3\epsilon}(H)|$ vertices, if a vertex $v \in V$ neighbors at least $(1 - \epsilon)|H|$ vertices in H then v has at least $(1 - \epsilon)|C|$ neighbors in C.*

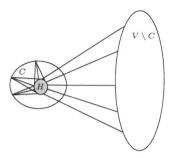

Fig. 1. An illustration of a Jellyfish Graph

3. *For all but $\epsilon|H|$ vertices in the graph, if $deg(v) \geq d$ then $v \in H$.*
4. $|\Gamma_{3\epsilon}(H)|/|C| \leq c.$

Intuitively, Item 1 of Definition 4 describes the fact that the vertices in C have many neighbors in H. Item 2 describes the fact that vertices that have many neighbors in H must have many neighbors in C too (so that the neighborhood relation to H is in a sense "representative" of the neighborhood relation to C). Item 3 describes the fact that most of the high-degree vertices in the graph are in H. Item 4 describes the fact that most vertices that neighbor most of H, are in C. Thus items 1 to 4 describe the dense-core as a dense set of vertices that contains most of the very high-degree nodes in the graph, neighbors most of these high-degree nodes, and are almost all the vertices in the graph that neighbor most of these high-degree nodes. Figure 1 shows an illustration of a Jellyfish graph.

Two notes are in place:

1. These assumptions hold for the AS-graph according to [30] for the values of k, d, c, ϵ we use.
2. Item 3 in Definition 4 will be relaxed in the full version of the paper [20] for $|H| = O(\log k)$ so that in this item ϵ will be any constant (that might be larger than 1).

3   The JellyCore Algorithm for Finding a Dense-Core in Jellyfish Graphs

In this section we describe a randomized algorithm that, given a graph $G = (V, E)$ that contains a (k, d, c, ϵ)-Jellyfish subgraph, finds a $(k, (8 \cdot c + 1)\sqrt{\epsilon})$-dense-core $\widehat{C}$ and an approximation of the nucleus H. Our algorithm and its analysis take some ideas from the Approximate-Clique Finding Algorithm of [19] (which is designed for dense graphs).

The algorithm is given query access to the graph G, and takes as input: k (the requested dense-core size), d (the minimal degree for nodes in the nucleus), ϵ, and a sample size s.

Algorithm 1. (The JellyCore algorithm for approximating C and H)

1. *Uniformly and independently at random select s vertices. Let S be the set of vertices selected.*
2. *Compute $\widehat{H} = \{v \in \Gamma(S) \mid deg(v) \geq d\}$. If $\widehat{H} = \emptyset$ then abort.*
3. *Compute the set $\Gamma_{2\epsilon}(\widehat{H})$ of vertices that neighbor all but at most of $2\epsilon|\widehat{H}|$ vertices in $\widehat{H}$.*
4. *Order the vertices in $\Gamma_{2\epsilon}(\widehat{H})$ according to their degree in the subgraph induced by $\Gamma_{2\epsilon}(\widehat{H})$ (breaking ties arbitrarily). Let $\widehat{C}$ be the first k vertices according to this order.*
5. *Return $\widehat{C}, \widehat{H}$*

Our main result is the following:

Theorem 1. *Let $G = (V, E)$ be a sparse graph that contains a (k, d, c, ϵ)-Jellyfish subgraph. Then, for $s \geq c'(n/d)\ln(|H| + 2)$, where c' is a constant, with probability at least $1 - e^{-(c'-1)}$, Algorithm 1 finds a set $\widehat{C}$ of size $|\widehat{C}| = k$ that is $O(\sqrt{\epsilon})$ close to being a clique, and finds a set $\widehat{H}$ that is a superset of H s.t. $|\widehat{H}| \leq (1 + \epsilon)|H|$. The time complexity of Algorithm 1 is $O(n \log n)$.*

Intuitively, the algorithm works in graphs that contain (k, d, c, ϵ)-Jellyfish subgraphs since in such graphs it suffices to sample a small set of vertices and observe their neighbors. The set of the neighbors with degree at least d is close to a nucleus H. In addition, in graphs that contain (k, d, c, ϵ)-Jellyfish subgraphs each vertex in C neighbors most of the vertices in H, and there might be only few vertices outside C that neighbor most of the vertices in H. Therefore, by taking the vertices that neighbor most of the vertices in H we get an approximation of C. However, in general graphs, if we sample a small set of vertices, the set of their neighbors might be a small random subset, so we won't be able to get any approximation of C.

We prove Theorem 1 by proving several lemmas. The lemmas and their proofs appear in the full version of the paper [20].

Assume for now that we have access to a superset $\overline{U}$ of H that contains vertices with degree at least d.

We next state our main lemma.

Lemma 1. *Suppose we order the vertices in $\Gamma_{2\epsilon}(\overline{U})$ according to their degree in the subgraph induced by $\Gamma_{2\epsilon}(\overline{U})$ (breaking ties arbitrarily). Let $\widehat{C}$ be the first k vertices according to this order. Then $\widehat{C}$ is $O(\sqrt{\epsilon})$ close to being a clique.*

Since we don't actually have access to a superset $\overline{U}$ of H that contains vertices with degree at least d, we sample the graph in order to get w.h.p. such a set.[4]

---

[4] We note that it is possible to search the graph for the vertices with degree at least d in linear time, which would not change (asymptotically) the running time of Algorithm 1. However, we shall need to perform random sampling in our sublinear algorithm, which is based on Algorithm 1, and hence we choose to introduce sampling at this stage. Furthermore, as we see in our implementation, in practice, we gain from using random sampling even when running Algorithm 1.

Specifically, we select s vertices uniformly and independently, where s should be at least $c'(n/d)\ln(|H|+2)$ for a constant c', and let S denote the subset of sampled vertices. Let $\widehat{H} = \{v \in \Gamma(S)|deg(v) \geq d\}$. Then

Lemma 2. *With probability at least* $1 - e^{-(c'-1)}$ *it holds that* $H \subseteq \widehat{H}$.

Proof of Theorem 1. The correctness of Algorithm 1 follows from Lemmas 1 and 2. It remains to compute the time complexity of the algorithm:

1. Steps 1 and 2: The most expensive operation is computing $\widehat{H}$. $\widehat{H}$ is computed by going over all the vertices in S, and adding the neighbors of each vertex with degree at least d to a list. (Thus a vertex can appear several times in the list). The time complexity is $\sum_{v \in S} deg(v) \leq \min\{2|E|, |S| \cdot n\} = O(n)$.

2. Step 3: This step is preformed in the following manner. First the multiset $\Gamma(\widehat{H})$ is computed, and then $\Gamma_{2\epsilon}(\widehat{H})$ is computed.

   (a) $\Gamma(\widehat{H})$ is computed by going over all the vertices in $\widehat{H}$, and adding the neighbors of each vertex to a list. (Here too a vertex can appear several times in the list). The time complexity is $\sum_{v \in \widehat{H}} deg(v) \leq \min\{2|E|, |\widehat{H}| \cdot n\} = O(n)$.

   (b) $\Gamma_{2\epsilon}(\widehat{H})$ is computed by the following algorithm: (i) Sort the vertices in the multiset $\Gamma(\widehat{H})$ according to the names of the vertices. (ii) For each vertex in $\Gamma(\widehat{H})$ count the number of times it appears in $\Gamma(\widehat{H})$. If it appears at least $(1 - 2\epsilon)|\Gamma(\widehat{H})|$ times then add the vertex to $\Gamma_{2\epsilon}(\widehat{H})$. The time complexity is $|\Gamma(\widehat{H})| \log |\Gamma(\widehat{H})| = O(n \log n)$.

3. Step 4: $\widehat{C}$ is computed by first computing the degrees in $\Gamma_{2\epsilon}(\widehat{H})$ of the vertices in $\Gamma_{2\epsilon}(\widehat{H})$, and then sorting the vertices in $\Gamma_{2\epsilon}(\widehat{H})$ according to this degree. Computing the degrees is upper bounded by $\sum_{v \in \Gamma_{2\epsilon}(\widehat{H})} deg(v) = \min\{2|E|, n \cdot |\Gamma_{2\epsilon}(\widehat{H})|\} = O(n)$. Therefore the time complexity of this step is upper bounded by $O(n) + |\Gamma_{2\epsilon}(\widehat{H})| \log(|\Gamma_{2\epsilon}(\widehat{H})|) \leq O(n) + O(k \log k) = O(n \log n)$.

Thus the time complexity of the algorithm is $O(n \log n)$.

4   A Sublinear Algorithm

In this section we modify the algorithm described in the previous section to get a sublinear algorithm that works under an additional assumption. For the sake of simplicity, we continue using the term Jellyfish subgraph, where we only add an additional parameter to its definition. Specifically, we say that a graph $G = (V, E)$ contains a (k, d, d', c, ϵ)-Jellyfish subgraph if it contains a (k, d, c, ϵ)-Jellyfish subgraph as described in Definition 4, and there are at most $\epsilon|H|$ vertices in the graph with degree larger than d'.

The next claim follows directly from a simple counting argument.

Claim 3. *Let* $G = (V, E)$ *be a sparse graph, where* $|E| \leq c''n$. *Then for any choice of* d, *the graph* G *contains at most* $d/2$ *vertices with degree larger than* $4c''n/d$.

Let H' be a subset of H that contains only vertices with degree at most d'. By the definition of a (k, d, d', c, ϵ)-Jellyfish subgraph, it holds that

$$|H| \geq |H'| \geq |H| - \epsilon|H| = (1 - \epsilon)|H|.$$

The algorithm is given query access to the graph $G = (V, E)$, and takes as input: k (the requested dense-core size), d (the minimal degree for nodes in the nucleus), d' (the high-degree threshold), ϵ, c'' (where $|E| \leq c''n$) and a sample size s.

Algorithm 2. (An algorithm for approximating C and H)

1. *Uniformly and independently at random select s vertices. Let S be the set of vertices selected.*
2. *Compute $S' = \{v \in S \mid deg(v) \leq 4c''n/d\}$.*
3. *Compute $\widehat{H}' = \{v \in \Gamma(S') \mid d \leq deg(v) \leq d'\}$. If $\widehat{H}' = \emptyset$ then abort.*
4. *Compute $\Gamma_{4\epsilon}(\widehat{H}')$ the set of vertices that neighbor all but at most $4\epsilon|\widehat{H}'|$ vertices in $\widehat{H}'$.*
5. *Compute $\widehat{C} = \{u \in \Gamma_{4\epsilon}(\widehat{H}')|deg(u) \geq d\}$.*
6. *Order the vertices in $\Gamma_{4\epsilon}(\widehat{H}')\backslash\widehat{C}$ according to their degree in the subgraph induced by $\Gamma_{4\epsilon}(\widehat{H}')$ (breaking ties arbitrarily). Let C'' be the first $k - |\widehat{C}|$ vertices according to this order.*
7. *$\widehat{C} \leftarrow \widehat{C} \cup C''$.*
8. *Return $\widehat{C}, \widehat{H}'$*

Our main result is the following:

Theorem 2. *Let $G = (V, E)$ be a sparse graph that contains a $(k, \Omega(n^{1-\beta}), O(n^{1-\beta/2}), c, \epsilon)$-Jellyfish subgraph. Then, for $s \geq c'(n/d)\ln(|H| + 2)$, where c' is a constant, with probability at least $1 - e^{1-c'/2}$ Algorithm 2 finds a set $\widehat{C}$ of size $|\widehat{C}| = k$ that is $O(\sqrt{\epsilon})$ close to being a clique, and finds a set $\widehat{H}'$ that is a superset of H' s.t. $|\widehat{H}'| \leq (1 + \epsilon)|H|$.[5] For $k = O(\log n)$ and $\beta \leq 2/5$, the time complexity of Algorithm 2 is $\widetilde{O}(n^{1-\beta/2})$[6].*

The proof of Theorem 2 appears in the full version of the paper [20].

5   Implementation

To demonstrate the usefulness of our algorithms beyond their theoretical contribution, we conducted a performance evaluation of our algorithm in comparison with the GreedyMaxClique algorithm of Siganos et al. [31] and the kCore algorithm of Carmi et al. [10] on real AS-graph data.

For our own algorithm we implemented the basic Algorithm 1 of Section 3. We did not implement the sublinear algorithm of Section 4. The AS graph contains only a

---

[5] Recall that $|H'| \geq (1 - \epsilon)|H|$, so Algorithm 2 indeed approximates H.

[6] The notation $\widetilde{O}(g(k))$ for a function g of a parameter k means $O(g(k) \cdot \text{polylog}(g(k)))$ where $\text{polylog}(g(k)) = \log^c(g(k))$ for some constant c.

handful of very high degree vertices, so the main assumption of Section 4 holds anyway. This means that the refinements of the sublinear algorithm, which ensure that we do not process too many such vertices, would not bring significant gains. Moreover, the basic JellyCore algorithm gave us excellent running times (see below), so we opted for simplicity and ease of programming.

All three algorithms were implemented in Java, using Sun's Java 5, using the open source library JUNG [23] (Java Universal Network/Graph Framework). We ran the algorithms on a 3GHz 4x multiprocessor Intel Xeon server with 4GB RAM, running RedHat Linux kernel 2.6.9.

We tested the algorithms on AS graphs constructed from data collected by the DIMES project [30]. DIMES is a large-scale distributed measurements effort that measures and tracks the evolution of the Internet from hundreds of different view-points, and provides detailed Internet topology graphs. We merged AS graphs from consecutive weeks starting from the first week of 2006 until reaching a total of 64 weeks in February 2007. This resulted in AS graphs that have a vertex count ranging from 11,000 to around 21,000 ASes.

All three algorithms accept the Internet AS graph as an input. The kCore algorithm used a degree of 29 (i.e., it produced a core in which the minimal residual node degree is 30). The parameters for our JellyCore Algorithm 1 were set as follows: Given the number of vertices n, we used a minimal nucleus degree of $d = n^{0.7}$, which gave $675 < d < 1100$ for our values of n. We picked $\epsilon = 0.1$ since we knew from earlier work that the AS graph contains a clique of 10–13 vertices—a smaller value of ϵ would have been essentially meaningless. The sample size s was calculated as follows: $s = 10 \cdot n^{0.3} \cdot \ln(3 \log(5 \log n))$ (this gave values $473 < s < 577$ for our values of n). To allow a fair comparison with the kCore algorithm, we set the required dense-core size k to be the exact core size returned by the kCore algorithm in each run ($67 < k < 91$ in all cases).

Since the JellyCore algorithm is randomized, we ran it 10 times on each input graph, each time with independent random samples. Each point plotted in the figures represents the average of these 10 runs.

5.1 Accuracy of the JellyCore Algorithm

Figure 2 (left) shows the percentage of matching vertices of JellyCore and kCore. In other words, if kCore returned a core Z and Jellycore returned a core J then Figure 2 (left) shows $100 \cdot |J \cap Z|/|Z|$. We can see that in all cases, between 92% and 95% of the core J returned by JellyCore is also in Z. Thus the results of JellyCore and kCore are very similar on the AS graph.

The figure also shows the percentage of matching vertices between the clique Q returned by GreedyMaxClique and the nucleus $\widehat{H}$ (here denoted by U). We can see that $\widehat{H}$ contains between 68% and 94% of the vertices of Q - and that this percentage improves as the number of vertices grows. Furthermore, we found by inspection that the JellyCore's J always completely includes the GreedyMaxClique Q.

Figure 2 (right) shows the density of the cores returned by JellyCore and of kCore as a function of the number of vertices of the graph. We can see that both densities are

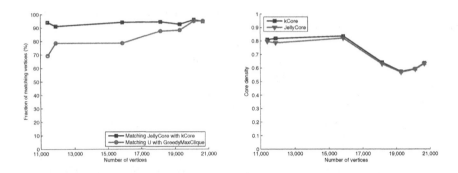

Fig. 2. Left: Percentage of matching vertices between JellyCore and kCore, and percentage of matching vertices between of $U (= \widehat{H})$ and GreedyMaxClique, for increasing graph sizes. Right: Core Density.

almost identical, particularly for $n \geq 18,000$ vertices. The density of GreedyMax-Clique is obviously 1, by the algorithm definition.

We can conclude that the practical results of the JellyCore algorithm, on the real AS graph, agree extremely well with the results of both kCore and GreedyMaxClique.

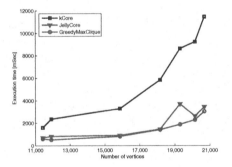

Fig. 3. Executions times

5.2   Execution Times

Figure 3 shows that the running times of JellyCore and GreedyMaxClique are almost identical, and that kCore is indeed slower: Jellycore runs about 6 times faster than kCore on the largest AS graphs. Moreover, the running time of kCore increases substantially as the number of vertices in the graph grows, while the growth in the running times of JellyCore and GreedyMaxClique is relatively minor.

Therefore, we can see that the JellyCore algorithm produces cores that are very similar to those kCore—at a fraction of the running time. In addition, JellyCore returns "for free" the nucleus $\widehat{H}$, which is essentially the clique Q discovered by GreedyMaxClique.

6 Conclusions

In this work we presented first a simple algorithm (JellyCore), and then a *sublinear* algorithm, for approximating the dense-core of a Jellyfish graph. We mathematically proved the correctness of our algorithms, under mild assumptions that hold for the AS graph. In our analysis we bounded the density of the cores our algorithms return, and analyzed their running time.

We also implemented our JellyCore algorithm and tested it on real AS-graph data. Our results show that the dense-core returned by JellyCore is very similar to the kCore of Carmi et al. [10], at a fraction of the running time, and the improvement is more prominent as the number of vertices increases. In addition, as a side effect JellyCore also approximates the clique returned by GreedyMaxClique of Siganos et al. [31].

Therefore, we have demonstrated that our randomized approach provides both a theoretically successful algorithm (with a rigorous asymptotic analysis of the discovered density and success probability)—and a successful practical algorithm.

References

1. Albert, R., Barabási, A.-L.: Topology of evolving networks: Local events and universality. Physical Review Letters 85(24), 5234–5237 (2000)
2. Alvarez-Hamelin, I., Dall'Asta, L., Barrat, A., Vespignani, A.: Large scale networks fingerprinting and visualization using the k-core decomposition. Proc. Neural Information Processing Systems (August 2005)
3. Asahiro, Y., Iwama, K., Tamaki, H., Tokuyama, T.: Greedily finding a dense subgraph. Journal of Algorithms 34, 203–221 (2000)
4. Bar, S., Gonen, M., Wool, A.: An incremental super-linear preferential Internet topology model. In: Barakat, C., Pratt, I. (eds.) PAM 2004. LNCS, vol. 3015, pp. 53–62. Springer, Heidelberg (2004)
5. Bar, S., Gonen, M., Wool, A.: A geographic directed preferential Internet topology model. Computer Networks 51(14), 4174–4188 (2007)
6. Barford, P., Bestavros, A., Byers, J., Crovella, M.: On the marginal utility of network topology measurements. In: Proc. ACM SIGCOMM (2001)
7. Bianconi, G., Barabási, A.L.: Competition and multiscaling in evolving networks. Europhysics Letters 54(4), 436–442 (2001)
8. Brunet, R.X., Sokolov, I.M.: Evolving networks with disadvantaged long-range connections. Physical Review E 66(026118) (2002)
9. Bu, T., Towsley, D.: On distinguishing between Internet power-law generators. In: Proc. IEEE INFOCOM 2002, New-York (April 2002)
10. Carmi, S., Havlin, S., Kirkpatrick, S., Shavitt, Y., Shir, E.: Medusa - new model of Internet topology using k-shell decomposition. Technical Report arXiv:cond-mat/0601240v1 (2006)
11. Carmi, S., Havlin, S., Kirkpatrick, S., Shavitt, Y., Shir, E.: A model of internet topology using k-shell decomposition. PNAS 2007. Proceedings of the National Academy of Sciences, USA 104(27), 11150–11154 (July 3, 2007)
12. Charikar, M.: Greedy approximation algorithms for finding dense components in graphs. In: Proc. APPROX (2000)
13. Chen, Q., Chang, H., Govindan, R., Jamin, S., Shenker, S., Willinger, W.: The origin of power laws in Internet topologies revisited. In: Proc. IEEE INFOCOM 2002, New-York (April 2002)

14. Faloutsos, C., Faloutsos, M., Faloutsos, P.: On power-law relationships of the Internet topology. In: Proc. of ACM SIGCOMM 1999, pp. 251–260 (August 1999)
15. Feige, U., Kortsarz, G., Peleg, D.: The dense k-subgraph problem. Algorithmica 29(3), 410–421 (2001)
16. Feige, U., Langberg, M.: Approximation algorithms for maximization problems arising in graph partitioning. Journal of Algorithms 41, 174–211 (2001)
17. Feige, U., Seltser, M.: On the densest k-subgraph problem. Technical report, Department of Applied Mathematics and Computer Science, The Weizmann Institute, Rehovot (1997)
18. Ge, Z., Figueiredo, D.R., Jaiswal, S., Gao, L.: On the hierarchical structure of the logical Internet graph. In: SPIE ITCOM (August 2001)
19. Goldreich, O., Goldwasser, S., Ron, D.: Property testing and its connections to learning and approximation. J. ACM 45, 653–750 (1998)
20. Gonen, M., Ron, D., Weinsberg, U., Wool, A.: Finding a dense-core in jellyfish graphs. Technical report, School of Electrical Enjeneering, Tel-Aviv University (2007)
21. Govindan, R., Tangmunarunki, H.: Heuristics for Internet map discovery. In: Proc. IEEE INFOCOM 2000, Tel-Aviv, Israel, pp. 1371–1380 (March 2000)
22. Håstad, J.: Clique is hard to approximate within $n^{1-\epsilon}$. Acta Mathematica 182, 105–142 (1999)
23. JUNG - the java universal network/graph framework (2007), http://jung.sourceforge.net/
24. Krapivsky, P.L., Rodgers, G.J., Render, S.: Degree distributions of growing networks. Physical Review Letters 86(5401) (2001)
25. Lakhina, A., Byers, J.W., Crovella, M., Xie, P.: Sampling biases in IP topology measurments. In: Proc. IEEE INFOCOM 2003 (2003)
26. Li, X., Chen, G.: A local-world evolving network model. Physica A 328, 274–286 (2003)
27. Mishra, N., Ron, D., Swaminathan, R.: A new conceptual clustering framework. Machine Learning 56, 115–151 (2004)
28. Reittu, H., Norros, I.: On the power law random graph model of the Internet. Performance Evaluation 55 (January 2004)
29. Sagie, G., Wool, A.: A clustering approach for exploring the Internet structure. In: Proc. 23rd IEEE Convention of Electrical & Electronics Engineers in Israel (IEEEI) (2004)
30. Shavitt, Y., Shir, E.: DIMES: Let the Internet measure itself. In: Proc. ACM SIGCOMM, pp. 71–74 (2005)
31. Siganos, G., Tauro, S.L., Faloutsos, M.: Jellyfish: A conceptual model for the as Internet topology. Journal of Communications and Networks (2006)
32. Subramanian, L., Agarwal, S., Rexford, J., Katz, R.H.: Characterizing the Internet hierarchy from multiple vantage points. In: Proc. IEEE INFOCOM 2002, New-York (April 2002)
33. Tangmunarunkit, H., Govindan, R., Jamin, S., Shenker, S., Willinger, W.: Network topology generators: Degree based vs. structural. In: Proc. ACM SIGCOMM (2002)
34. Tauro, L., Palmer, C., Siganos, G., Faloutsos, M.: A simple conceptual model for Internet topology. In: IEEE Global Internet, San Antonio, TX (November 2001)
35. Willinger, W., Govindan, R., Jamin, S., Paxson, V., Shenker, S.: Scaling phenomena in the Internet: Critically examining criticality. Proceedings of the National Academy of Sciences of the United States of America 99, 2573–2580 (February 2002)
36. Winick, J., Jamin, S.: Inet-3.0: Internet topology generator. Technical Report UM-CSE-TR-456-02, Department of EECS, University of Michigan (2002)

A Geometric Preferential Attachment Model of Networks II

Abraham D. Flaxman, Alan M. Frieze*, and Juan Vera

Department of Mathematical Sciences,
Carnegie Mellon University,
Pittsburgh PA15213,
U.S.A.

Abstract. A detailed understanding of expansion in complex networks can greatly aid in the design and analysis of algorithms for a variety of important network tasks, including routing messages, ranking nodes, and compressing graphs. This has motivated several recent investigations of expansion properties in real-world graphs and also in random models of real-world graphs, like the preferential attachment graph. The results point to a gap between real-world observations and theoretical models. Some real-world graphs are expanders and others are not, but a graph generated by the preferential attachment model is an expander **whp**.

We study a random graph G_n that combines certain aspects of geometric random graphs and preferential attachment graphs. This model yields a graph with power-law degree distribution where the expansion property depends on a tunable parameter of the model.

The vertices of G_n are n sequentially generated points $x_1, x_2, \ldots, x_n$ chosen uniformly at random from the unit sphere in $\mathbf{R}^3$. After generating x_t, we randomly connect it to m points from those points in $x_1, x_2, \ldots, x_{t-1}$ ....

1 Introduction

During the last decade a large body of research has centered on understanding and modeling the structure of large-scale networks like the Internet and the World Wide Web. Several recent books provide a general introduction to this topic [37] and [40]. One important feature identified in early experimental studies (including [3,12,22]) is that the vertex degree distribution of many real-world networks has a heavy-tailed property, which may follow a power-law (i.e., the proportion of vertices of degree at least k is proportional to $k^{-\alpha}$ for some constant α). This has driven the investigation of random graph distributions which generate heavy-tailed degree distributions, including the fixed degree sequence model, the copying model, and the preferential attachment model.

The preferential attachment model and its derivatives have been particularly popular for theoretical analysis. Preferential attachment was proposed as a model for real-world complex networks by Barabási and Albert [4]. The distribution

---

* Supported in part by NSF grant CCF-200945.

A. Bonato and F.R.K. Chung (Eds.): WAW 2007, LNCS 4863, pp. 41–55, 2007.
© Springer-Verlag Berlin Heidelberg 2007

was formalized by Bollobás and Riordan [9], and in [11] it was proved rigorously that **whp** a graph chosen according to this distribution has a power-law degree distribution with complementary cumulative distribution function (ccdf) $\Pr[\deg(v) \geq k] = \Theta(k^{-2})$. By changing the initial attactiveness or incorporating more random addition and deletion, the power of the ccdf power-law can be tuned to take any value in the interval $(1, \infty)$ [13,17].

However, there are some significant differences between graphs generated by preferential attachment and those found in the real world. One major difference is found in their expansion properties. Mihail, Papadimitriou, and Saberi [35] showed that **whp** the preferential attachment model has conductance bounded below by a constant. On the other hand, Blandford, Blelloch and Kash [7] found that some WWW related graphs have smaller separators than the preferential attachment model predicts. This observation is consistent with observations due to Estrada [19], who found that half of the real-world networks he looked at were good expanders and the other half were not so good. The perturbed random graph framework provides one approach to understanding expansion in real-world networks [23], but it does not give a generative procedure. This paper investigates a generative procedure, based on a geometric modification of the preferential attachment model, which yields a graph that might or might not be a good expander, depending on a tunable parameter of the geometry. This is a strict generalization of the geometric preferential attachment graph developed in [24] which was designed specifically to avoid being a good expander.

The primary contribution of this paper is to provide a parameterised model that exhibits a sharp transition between low and high conductance. Choosing this parameter appropriately provides a unified approach to generating preferential attachment graphs with and without good expansion processes.

1.1   The Random Process

In [24] we studied a process which generates a sequence of graphs $G_t, t = 1, 2, \ldots, n$. The graph $G_t = (V_t, E_t)$ has t vertices and mt edges. Here V_t is a subset of t random points on S, the surface of the sphere in $\boldsymbol{R}^3$ of radius $\frac{1}{2\sqrt{\pi}}$ (so that $area(S) = 1$). After randomly choosing $x_{t+1} \in S$, it is connected, by preferential attachment (i.e. proportional to degree), to m vertices in V_t among those of distance at most r from x_{t+1}. We showed that this graph has a power law degree distribution, small seperators and a moderate diameter. In this paper we provide a "smoothed" version of this model, instead of choosing proportional to degree among those vertices within distance r of x_{t+1}, the m neighbors of x_t are chosen proportional to degree and some function of the distance to x_{t+1}.

Let $F : \boldsymbol{R}_+ \to \boldsymbol{R}_+$. Define

$$I = \int_S F(|u - u_0|)du = \frac{1}{2} \int_{x=0}^{\pi} F(x) \sin x dx)$$

where u_0 is any point in S and $0 \leq |u - u_0| \leq \pi$ is the angular distance from u to u_0 along a great circle. Other parameters of the process are $m > 0$ the number of edges added in every step and $\alpha \geq 0$ a measure of the bias towards self loops.

- **Time step** 0: To initialize the process, we start with G_0 being the Empty Graph.
- **Time step** $t+1$: We choose vertex x_{t+1} uniformly at random in S and add it to G_t. Let

$$T(x_{t+1}) = \sum_{v \in V_t} F(|x_{t+1} - v|) \deg_t(v).$$

We add m random edges (x_{t+1}, y_i), $i = 1, 2, \ldots, m$ incident with x_{t+1}. Here, each y_i is chosen independently from $V_{t+1} = V_t \cup \{x_{t+1}\}$ (parallel edges and loops are permitted), such that for each $i = 1, \ldots, m$, for all $v \in V_t$,

$$\mathbf{Pr}(y_i = v) = \frac{\deg_t(v) F(|x_{t+1} - v|)}{\max\left(T(x_{t+1}), \alpha m It\right)}$$

and

$$\mathbf{Pr}(y_i = x_{t+1}) = 1 - \frac{T(x_{t+1})}{\max\left(T(x_{t+1}), \alpha m It\right)}$$

(When $t = 0$ we have $\mathbf{Pr}(y_i = x_1) = 1$.)

For $z > 0$ we define

$$I_z = \frac{1}{2} \int_{x=0}^{z} F(x) \sin x \, dx \text{ and } J_z = I - I_z.$$

Where possible we will illustrate our theorems using the canonical functions:

$$F_0(u) = 1_{|u| \leq r}, \qquad\qquad r \geq n^{\epsilon - 1/2}.$$

$$F_1(u) = \frac{1}{\max\{n^{-\delta}, u\}^{\beta}} \qquad where \ \delta < 1/2.$$

$$F_2(u) = e^{-\beta u} \qquad\qquad \beta = \beta(n) \geq 0.$$

Notice that F_0 corresponds to the model presented in [24]. Also notice that without the $n^{-\delta}$ term in the definition of F_1 for $\beta \geq 2$ we would have $I = \infty$. One can justify its inclusion (for some value of δ) from the fact that **whp** the minimum distance between the points in V_n is greater than $1/n \ln n$.

Observe that

$$I_z(F_0) = \frac{1}{2}(1 - \cos(\min\{z, r\})).$$

$$I_z(F_1) = \begin{cases} \frac{\beta n^{\delta(\beta-2)}}{4(\beta-2)} + O(n^{(\beta-4)\delta} + z^{2-\beta}) & z \geq n^{-\delta}, \beta > 2. \\ \Theta(z^{2-\beta}) + O(n^{(\beta-2)\delta}) & z \geq n^{-\delta}, \beta < 2 \\ \ln(n^{\delta} z) + O(1) & z \geq n^{-\delta}, \beta = 2 \end{cases}$$

$$I_z(F_2) = \frac{1}{2(1 + \beta^2)}(1 - e^{-\beta z}(\cos z + \beta \sin z)).$$

Let $d_k(t)$ denote the number of vertices of degree k at time t and let $\overline{d}_k(t)$ denote the expectation of $d_k(t)$.

We will first prove the following result about the degree distribution and the existence of small separators:

Theorem 1

(a) *Suppose that $\alpha > 2$ and in addition that*

$$\int_{x=0}^{\pi} F(x)^2 \sin x \, dx = O(n^\theta I^2) \tag{1}$$

where $\theta < 1$ is a constant.

    Then there exists a constant $\gamma_1 = \gamma_1(\alpha, \theta) > 0$ such that for all $k = k(n) \geq m$,

$$\bar{d}_k(n) = e^{\varphi_k(m,\alpha)} \left(\frac{m}{k}\right)^{1+\alpha} n + O(n^{1-\gamma_1}) \tag{2}$$

where $\varphi_k(m, \alpha) = O(1)$ tends to a constant $\varphi_\infty(m, \alpha)$ as $k \to \infty$.

    Furthermore, for n sufficiently large, the random variable $d_k(n)$ satisfies the following concentration inequality: Let $\zeta > 0$ be constant.

$$\mathbf{Pr}(|d_k(n) - \bar{d}_k(n)| \geq I^2 n^{\max\{1/2, 2/\alpha\}+\zeta}) \leq e^{-n^\zeta}. \tag{3}$$

(b) *Suppose that $\alpha > 0$ and $m_0 \leq m$ where m_0 is a sufficiently large constant and $\varphi, \eta = o(1)$ are such that $\eta n \to \infty$ and $J_\eta \leq \varphi I$. Then* **whp***, V_n can be partitioned into $T, \bar{T}$ such that $|T|, |\bar{T}| \sim n/2$, and there are $\tilde{O}((\eta + \varphi)n)$ edges between T and $\bar{T}$.*

Remark 1. Note that the exponent in (a) does not depend on the particular function F. F manifests itself only through the error terms.

For Part (a) of the above theorem:

$F = F_0$: $\theta = 0$.
$F = F_1$, $\beta > 2$: $\theta = 2\delta$.
$F = F_1$, $\beta < 2$: $\theta = 0$.
$F = F_1$, $\beta = 2$: $\theta = 2\delta$.
$F = F_2$: $\theta = 0$.

    For Part (b) of the above theorem:

$F = F_0$: $\eta = r$, $\varphi = 0$.
$F = F_1$, $\beta > 2$: $\eta = n^{-\delta}$, $\varphi = O(n^{-(\beta-2)\delta})$.
$F = F_1$, $\beta = 2$: $\eta = \frac{\ln \ln n}{\ln n}$, $\varphi = O(\eta)$.

    We now consider the connectivity and diameter of G_n. For this we will place some more restrictions on F.

    Define the parameter $\rho(\mu)$ by

$$I_\rho = \mu I. \tag{4}$$

    We will say that F is *smooth* (for some value of μ) if

(S1) F is monotone non-increasing.
(S2) $\rho^2 n \geq L \ln n$ for some sufficiently large constant L.
(S3) $\rho^2 F(2\rho) \geq c_3 I$ for some c_3 which is bounded away from zero.

Theorem 2. *Suppose that $\alpha > 2$ and F is smooth for some constant $\mu > 0$ and $m \geq K \ln n$ for K sufficiently large. Then* **whp**

(a) G_n *is connected.*
(b) G_n *has diameter $O(\ln n/\rho)$.*

For the above theorem:

$F = F_0$: $I \sim r^2/4$ and so we can take $\mu \sim 1/4$, $\rho = r/2$, $c_3 \sim 1$.
$F = F_1$, $\beta > 2$: $I \sim \frac{n^{\delta(\beta-2)}}{2(\beta-2)}$ and so we can take $\mu \sim 1/4$, $\rho = n^{-\delta}/2$, $c_3 \sim (\beta - 2)/2$.
$F = F_1$, $\beta < 2$: $I = \Theta(1)$ and we can take $\rho = 1$, $\mu = \Omega(1)$, $c_3 = \Omega(1)$.
$F = F_2$: $I = \Theta(1)$ and we can take $\rho = 1$, $\mu = \Omega(1)$, $c_3 = \Omega(1)$.

We have a problem fitting the case of F_1 with $\beta = 2$ into the theorem.
    We now consider conditons under which G_n is an expander.
    Let F be *tame* if there exist absolute constants C_1, C_2 such that

(T1) $F(x) \geq C_1$ for $0 < x \leq \pi$.
(T2) $I \leq C_2$.

We note that F_1 with $\beta < 2$ is tame since $F_1(x) \geq \pi^{-\beta}$ for $0 \leq \pi$ and

$$I = \frac{1}{2} \int_{x=0}^{\pi} x^{-\beta} \sin x \, dx \leq \frac{\pi^{2-\beta}}{2(2-\beta)}.$$

The *conductance* Φ of G_n is defined by

$$\Phi = \min_{\deg_n(K) \leq mn} \Phi(K) = \min_{\deg_n(K) \leq mn} \frac{|E(K : \bar{K})|}{\deg_n(K)}.$$

Theorem 3. *If $\alpha > 2$ and F is tame and $m \geq K \ln n$ for K sufficiently large then* **whp**

(a) G_n *has conductance bounded away from zero.*
(b) G_n *is connected.*
(c) G_n *has diameter $O(\log_m n)$.*

Mihail [31] has empirical results on the conductance of G_n in the case where $F = F_1$. They observe poor conductance when $\beta < 2$ and good conductance when $\beta > 2$. This fits nicely with the results of Theorems 2 and 3.

The role of α: This parameter was introduced in [24] as a means of overcoming a difficult technical problem. When $\alpha > 2$ it facilitates a proof of Lemma 2. On the positive side, it does give a parameter that effects the power law. On the negative side, when $\alpha > 2$, there will **whp** be isolated vertices, unless we make m grow at least as fast as $\ln n$. It is for us, an interesting open question, as to how to prove our results with $\alpha = 0$.

2 Outline of the Paper

We prove a likely power law for the degree sequence in Section 3. We follow a standard practise and prove a recurrence for the expected number of vertices of degree k at time step t. Unfortunatley, this involves the estimation of the expectation of the reciprocal of a random variable and to handle this, we show that this random variable is concentrated. This is quite technical and is done in Section 3.2. In Section 4 we show that under the assumptions of Theorem 1(b) there are small separators. This is relatively easy, since any give great circle can **whp** be used to define a small separator.

The proof of connectivity when m grows logarithmically with n is left to the full paper. The idea is to show that **whp** the sub-graph $G_n(B)$ induced by a ball B of radius ρ, centered in $u \in S$, is connected. and has small diameter. We then show that the union of the $G_n(B)$'s for $u = x_1, x_2, \ldots, x_n$ is connected and has small diameter.

3 Proving a Power Law

3.1 Establishing a Recurrence for $\overline{d}_k(t)$: The Expected Number of Vertices of Degree k at Time t

Our approach to proving Theorem 1(a) is to find a recurrence for $\overline{d}_k(t)$. For $k \in \mathbf{N}$ define $D_k(t) = \{v \in V_t : \deg_t(v) = k\}$. Thus $d_k(t) = |D_k(t)|$. Also, define $d_{m-1}(t) = 0$ and $\overline{d}_{m-1}(t) = 0$ for all integers t with $t > 0$. Let $\eta_k(G_t, x_{t+1})$ denote the (conditional) probability that a parallel edge from x_{t+1} to a vertex of degree no more than k is created at time $t + 1$. Then,

$$\eta_k(G_t, x_{t+1}) = O\left(\min\left\{\binom{m}{2} \sum_{i=m}^{k} \sum_{v \in D_i(t)} \frac{F(|x_{t+1} - v|)^2 \, i^2}{\max\{\alpha m I t, T(x_{t+1})\}^2}, 1\right\}\right). \quad (5)$$

Then for $k \geq m$,

$$\mathbf{E}\left[d_k(t+1) \mid G_t, x_{t+1}\right] = d_k(t)$$
$$+ m \sum_{v \in D_{k-1}(t)} \frac{(k-1)F(|x_{t+1} - v|)}{\max\{\alpha m I t, T(x_{t+1})\}} - m \sum_{v \in D_k(t)} \frac{kF(|x_{t+1} - v|)}{\max\{\alpha m I t, T(x_{t+1})\}}$$
$$+ \mathbf{Pr}\left[deg_{t+1}(x_{t+1}) = k) \mid G_t, x_{t+1}\right] + O(\eta_k(G_t, x_{t+1})). \quad (6)$$

Let $\mathcal{A}_t$ be the event

$$\{|T(x_{t+1}) - 2mIt| \leq C_1 I m t^\gamma \ln n\}$$

where

$$\max\{2/\alpha, \theta\} < \gamma < 1$$

and C_1 is some sufficiently large constant.

Note that if

$$t \geq t_0 = (\ln n)^{2/(1-\gamma)} \tag{7}$$

then

$$\mathcal{A}_t \text{ implies } T(x_{t+1}) \leq \alpha m I t.$$

Then, for $t \geq t_0$,

$$\mathbf{E}\left[\sum_{v \in D_k(t)} \frac{kF(|x_{t+1} - v|)}{\max\{\alpha m I t, T(x_{t+1})\}}\right]$$

$$= \mathbf{E}\left[\sum_{v \in D_k(t)} \frac{kF(|x_{t+1} - v|)}{\max\{\alpha m I t, T(x_{t+1})\}} \,\middle|\, \mathcal{A}_t\right] \mathbf{Pr}\left[\mathcal{A}_t\right]$$

$$+ \mathbf{E}\left[\sum_{v \in D_k(t)} \frac{kF(|x_{t+1} - v|)}{\max\{\alpha m I t, T(x_{t+1})\}} \,\middle|\, \neg\mathcal{A}_t\right] \mathbf{Pr}\left[\neg\mathcal{A}_t\right]$$

$$= \frac{k}{\alpha m t}\mathbf{E}\left[d_k(t)|\mathcal{A}_t\right]\mathbf{Pr}\left[\mathcal{A}_t\right] + O\left(1\right)\mathbf{Pr}\left[\neg\mathcal{A}_t\right]$$

$$= \frac{k\bar{d}_k(t)}{\alpha m t} - \frac{k}{\alpha m t}\mathbf{E}\left[d_k(t)|\neg\mathcal{A}_t\right]\mathbf{Pr}\left[\neg\mathcal{A}_t\right] + O\left(1\right)\mathbf{Pr}\left[\neg\mathcal{A}_t\right]$$

$$= \frac{k\bar{d}_k(t)}{\alpha m t} + O\left(k\right)\mathbf{Pr}\left[\neg\mathcal{A}_t\right]$$

In Lemma 2 below we prove that

$$\mathbf{Pr}\left[\neg\mathcal{A}_t\right] = O\left(n^{-2}\right). \tag{8}$$

Thus, if $t \geq t_0$ then

$$\mathbf{E}\left[\sum_{v \in D_k(t)} \frac{kF(|x_{t+1} - v|)}{\max\{\alpha m I t, T(x_{t+1})\}}\right] = \frac{k\bar{d}_k(t)}{\alpha m t} + O\left(k/n^2\right). \tag{9}$$

In a similar way

$$\mathbf{E}\left[\sum_{v \in D_{k-1}(t)} \frac{(k-1)F(|x_{t+1} - v|)}{\max\{\alpha m I t, T(x_{t+1})\}}\right] = \frac{(k-1)\bar{d}_{k-1}(t)}{\alpha m t} + O\left(k/n^2\right). \tag{10}$$

On the other hand, given G_t, x_{t+1}, if

$$p = 1 - \frac{T(x_{t+1})}{\max\left(T(x_{t+1}), \alpha m I t\right)}$$

then

$$\mathbf{Pr}\left[\deg_{t+1}(x_{t+1} = k) \mid G_t, x_{t+1}\right] = \mathbf{Pr}\left[\text{Bi}(m, p) = k - m\right]$$

So, if $t \geq t_0$,

$$\mathbf{Pr}\left[\deg_{t+1}(x_{t+1}=k)\right]=\binom{m}{k-m}\mathbf{E}\left[p^{k-m}(1-p)^{2m-k}\middle|\mathcal{A}_t\right]\mathbf{Pr}\left[\mathcal{A}_t\right]+O(\mathbf{Pr}\left[\neg\mathcal{A}_t\right])$$

$$=\binom{m}{k-m}\left(1-\frac{2}{\alpha}\right)^{k-m}\left(\frac{2}{\alpha}\right)^{2k-m}(1+O(t^{\gamma-1}\ln n))\mathbf{Pr}\left[\mathcal{A}_t\right]+O(n^{-2})$$

$$=\binom{m}{k-m}\left(1-\frac{2}{\alpha}\right)^{k-m}\left(\frac{2}{\alpha}\right)^{2k-m}+O(t^{\gamma-1}\ln n).$$

Now note that from equations (5) and (8) that if

$$t \geq t_1 = n^{(\gamma+\theta)/2\gamma}$$

and

$$k \leq k_0(t) = n^{(\gamma-\theta)/4}$$

then, from (1), we see that

$$\mathbf{E}(\eta_k(G_t, x_{t+1})) = O\left(\frac{k^2 n^\theta}{t}\right) = O(t^{\gamma-1}). \tag{11}$$

Taking expectations on both sides of (6) and using (9,10,11), we see that if $t \geq t_0$ and $k \leq k_0(t)$ then

$$\bar{d}_k(t+1) = \bar{d}_k(t) + \frac{k-1}{\alpha t}\bar{d}_{k-1}(t) - \frac{k}{\alpha t}\bar{d}_k(t)$$
$$+ \binom{m}{k-m}\left(1-\frac{2}{\alpha}\right)^{k-m}\left(\frac{2}{\alpha}\right)^{2m-k} + O\left(t^{\gamma-1}\ln n\right) \tag{12}$$

We consider the recurrence given by $f_{m-1} = 0$ and for $k \geq m$,

$$f_k = \frac{k-1}{\alpha}f_{k-1} - \frac{k}{\alpha}f_k + \binom{m}{k-m}\left(1-\frac{2}{\alpha}\right)^{k-m}\left(\frac{2}{\alpha}\right)^{2m-k}, \tag{13}$$

which, for $k > 2m$, has solution

$$f_k = f_{2m}\prod_{i=2m+1}^{k}\frac{i-1}{i+\alpha} = f_{2m}e^{\varphi_k(m,\alpha)}\left(\frac{m}{k}\right)^{\alpha+1}. \tag{14}$$

Here $\varphi_k(m,\alpha) = O(1)$ tends to a limit $\varphi_\infty(m,\alpha)$ depending only on m, α as $k \to \infty$. Furthermore, $\lim_{m\to\infty}\varphi_\infty(\alpha, m) = 0$. We also have

$$f_{m+i} \leq f_{2m}\prod_{j=i+1}^{m}\left(1 + \frac{\alpha+1}{m+j-1}\right) \leq e^{2\alpha+3}f_{2m}.$$

It follows that (14) is also valid for $m \leq k \leq 2m$ with $\varphi_k(m,\alpha) = O(1)$.

We finish the proof of (2) by showing that there exists a constant $M > 0$ such that

$$|\overline{d}_k(t) - f_k t| \leq M(n^{1-(\gamma-\theta)/4} + t^\gamma \ln n) \tag{15}$$

for all $0 \leq t \leq n$ and $m \leq k \leq k_0(t)$.

We have that (15) is trivially true for $t < t_1$, and for $t \geq t_1$ and $k > k_0(t)$ it follows from $\overline{d}_k(t) \leq 2mt/k$.

Now, let $\Theta_k(t) = \overline{d}_k(t) - f_k t$. Then for $t \geq t_1$ and $m \leq k \leq k_0(t)$,

$$\Theta_k(t+1) = \frac{k-1}{\alpha t}\Theta_{k-1}(t) - \frac{k}{\alpha t}\Theta_k(t) + O(t^{\gamma-1}\ln n). \tag{16}$$

Let L denote the hidden constant in $O(t^{\gamma-1}\ln n)$ of (16). Our inductive hypothesis $\mathcal{H}_t$ is that

$$|\Theta_k(t)| \leq M(t_1 + t^\gamma \ln n)$$

for every $m \leq k \leq k_0(t)$ and M sufficiently large. Assume that $t \geq t_1$. Then $k \ll t$ in the current range of interest, and so from (16),

$$|\Theta_k(t+1)| \leq M(t_1 + t^\gamma \ln n) + Lt^{\gamma-1}\ln n$$
$$\leq M(t_1 + (t+1)^\gamma \ln n).$$

This verifies $\mathcal{H}_{t+1}$ and completes the proof by induction.

3.2 Concentration of $T(u)$

Now we turn our attention to prove that $T(u)$ is concentrated around its mean.

Lemma 1. *Let $u \in S$ and $t > 0$ then $\mathbf{E}[T(u)] = 2Imt$.*

Proof

$$\mathbf{E}[T(u)] = \mathbf{E}\left[\sum_{v \in V_t} \deg_t(v)F(|u-v|)\right] = I\sum_{v \in V_t}\deg_t(v) = 2Imt. \qquad \square$$

Lemma 2. *If $t > 0$ and u is chosen randomly from S then*

$$\mathbf{Pr}\left[|T(u) - 2Imt| \geq mI(t^{2/\alpha} + t^{1/2}\ln t)\ln n\right] = O\left(n^{-2}\right).$$

Proof. We use the Azuma-Hoeffding inequality. One may be a little concerned here that our probability space is not discrete. Although it is not really necessary, one could replace S by 2^{2^n} randomly chosen points X and sample uniformly from these. Then **whp** the change in distribution would be negligible. With this re-assurance, fix τ, with $1 \leq \tau < t$. Fix G_τ and let $G_t = G_t(G_\tau, x_{\tau+1}, y_1, \ldots, y_m)$ and $\hat{G}_t = G_t(G_\tau, \hat{x}_{\tau+1}, \hat{y}_1, \ldots, \hat{y}_m)$, where $x_{\tau+1}, \hat{x}_{\tau+1} \in S$ and $y_1, \ldots, y_m, \hat{y}_1, \ldots, \hat{y}_m \in V_\tau$. We couple the construction of G_t and $\hat{G}_t$, starting at time step $\tau + 1$ with the graph G_τ and $\hat{G}_\tau$ respectively. Then, for every step $\sigma > \tau + 1$ we choose the same point $x_\sigma \in S$ in both and for every $i = 1, \ldots, m$ we choose $u_i, \hat{u}_i \in V_\sigma$ such that each marginal is the correct marginal and such that the probability of choosing the same vertex is maximized.

Notice that we have

$$\mathbf{Pr}\left[u_i = v = \hat{u}_i\right] = \min\left(\frac{\deg_{G_{\sigma-1}}(v)F(|v - x_\sigma|)}{\max\left(T(x_\sigma), \alpha m I(\sigma - 1)\right)}, \frac{\deg_{\hat{G}_{\sigma-1}}(v)F(|v - x_\sigma|)}{\max\left(\hat{T}(x_\sigma), \alpha m I(\sigma - 1)\right)}\right)$$

for every $v \in V_{\sigma-1}$. Also,

$$\mathbf{Pr}\left[u_i = x_\sigma = \hat{u}_i\right] = 1 - \max\left(\frac{T(x_\sigma)}{\max\left(T(x_\sigma), \alpha m I(\sigma-1)\right)}, \frac{\hat{T}(x_\sigma)}{\max\left(\hat{T}(x_\sigma), \alpha m I(\sigma-1)\right)}\right)$$

Now, for $u \in S$ let

$$\Delta_\sigma(u) := \Delta_{\sigma,\tau}(u) = \sum_{\rho=\tau}^{\sigma} \sum_{i=1}^{m} |F(|u - u_i^\rho|) - F(|u - \hat{u}_i^\rho|)|.$$

Lemma 3. *Let $t \geq 1$ and let u be a random point in S. Then for some constant $C > 0$,*

$$\mathbf{E}\left[\Delta_t(u)\right] \leq CmI\left(\frac{t}{\tau}\right)^{2/\alpha}.$$

Proof. We begin with

$$\mathbf{E}\left[|F(|w - u_i^\rho|) - F(|w - \hat{u}_i^\rho|)| \, \big| \, u_i^j, \hat{u}_i^j : i = 1, \ldots, m, \, j = 1, \ldots, \sigma\right] \leq 2I\mathbb{1}_{u_i^\rho \neq \hat{u}_i^\rho}.$$

Therefore if we define for every $\tau < \sigma \leq t$

$$\Delta_\sigma = \sum_{\rho=\tau}^{\sigma} \sum_{i=1}^{m} \mathbb{1}_{u_i^\sigma \neq \hat{u}_i^\sigma},$$

we have

$$\mathbf{E}\left[\Delta_\sigma(u)\right] \leq 2I\mathbf{E}\left[\Delta_\sigma\right].$$

Fix $\tau < \sigma \leq t$. We have then

$$\Delta_\sigma = \Delta_{\sigma-1} + \sum_{i=1}^{m} \mathbb{1}_{u_i^\sigma \neq \hat{u}_i^\sigma}. \tag{17}$$

Now fix $1 \le i \le m$. Taking expectations with respect to our coupling,

$$\mathbf{E}\left[1_{u_i^\sigma \ne \hat{u}_i^\sigma} | G_{\sigma-1}, \hat{G}_{\sigma-1}, x_\sigma\right] = 1 - \mathbf{Pr}\left[u_i^\sigma = \hat{u}_i^\sigma | G_{\sigma-1}, \hat{G}_{\sigma-1}, x_\sigma\right]$$

$$= \max\left(\frac{T(x_\sigma)}{\max\left(T(x_\sigma), \alpha m I(\sigma-1)\right)}, \frac{\hat{T}(x_\sigma)}{\max\left(\hat{T}(x_\sigma), \alpha m I(\sigma-1)\right)}\right)$$

$$- \sum_{v \in V_{\sigma-1}} \min\left(\frac{\deg_{G_{\sigma-1}}(v) F(|v - x_\sigma|)}{\max\left(T(x_\sigma), \alpha m I(\sigma-1)\right)}, \frac{\deg_{\hat{G}_{\sigma-1}}(v) F(|v - x_\sigma|)}{\max\left(\hat{T}(x_\sigma), \alpha m I(\sigma-1)\right)}\right)$$

$$\le \frac{\max\left(T(x_\sigma), \hat{T}(x_\sigma)\right) - \sum_{v \in V_{\sigma-1}} \min\left(\deg_{G_{\sigma-1}}(v), \deg_{\hat{G}_{\sigma-1}}(v)\right) F(|v - x_\sigma|)}{\max\left(T(x_\sigma), \hat{T}(x_\sigma), \alpha m I(\sigma-1)\right)}$$

$$\tag{18}$$

$$\le \frac{\sum_{v \in V_{\sigma-1}} |\deg_{G_{\sigma-1}}(v) - \deg_{\hat{G}_{\sigma-1}}(v)| F(|v - x_\sigma|)}{\max\left(T(x_\sigma), \hat{T}(x_\sigma), \alpha m I(\sigma-1)\right)}$$

$$\tag{19}$$

$$\le \frac{\sum_{v \in V_{\sigma-1}} |\deg_{G_{\sigma-1}}(v) - \deg_{\hat{G}_{\sigma-1}}(v)| F(|v - x_\sigma|)}{\alpha m I(\sigma-1)}$$

Inequality (18), follows from

$$\max\left(\frac{a}{\max(a,c)}, \frac{b}{\max(b,c)}\right) = \frac{\max(a,b)}{\max(a,b,c)}$$

and

$$\min\left(\frac{a}{b}, \frac{c}{d}\right) \ge \frac{\min(a,c)}{\max(b,d)}.$$

Inequality (19) is a consequence of $\max\{\sum_i a_i, \sum_i b_i\} - \sum_i \min\{a_i, b_i\} \le \sum_i |a_i - b_i|$.

Therefore

$$\mathbf{E}\left[\Delta_\sigma \,\middle|\, G_{\sigma-1}, \hat{G}_{\sigma-1}\right] \le \Delta_{\sigma-1} + \frac{\sum_{v \in V_{\sigma-1}} |\deg_{G_{\sigma-1}}(v) - \deg_{\hat{G}_{\sigma-1}}(v)|}{\alpha(\sigma-1)}. \tag{20}$$

But, for each $v \in V_{\sigma-1}$ we have

$$|\deg_{G_{\sigma-1}}(v) - \deg_{\hat{G}_{\sigma-1}}(v)| \le \sum_{j=\tau}^{\sigma-1} \sum_{i=1}^{m} (1_{u_i^j = v, \, \hat{u}_i^j \ne v} + 1_{u_i^j \ne v, \, \hat{u}_i^j = v})$$

and thus

$$\sum_{v \in V_{\sigma-1}} |\deg_{G_{\sigma-1}}(v) - \deg_{\hat{G}_{\sigma-1}}(v)| \le \sum_{j=\tau}^{\sigma-1} \sum_{i=1}^{m} \sum_{v \in V_{\sigma-1}} \left(1_{u_i^j = v, \, \hat{u}_i^j \ne v} + 1_{u_i^j \ne v, \, \hat{u}_i^j = v}\right) \le 2\Delta_{\sigma-1}.$$

Going back to (20) we have

$$\mathbf{E}\left[\Delta_\sigma\right] \le \mathbf{E}\left[\Delta_{\sigma-1}\right]\left(1 + \frac{2}{\alpha(\sigma-1)}\right),$$

so, $\mathbf{E}\left[\Delta_t\right] \le e^{O(1)}\left(\frac{t}{\tau}\right)^{2/\alpha}\mathbf{E}\left[\Delta_\tau\right]$. Now, $\Delta_\tau \le m$, because the graphs G_τ and $\hat{G}_\tau$ differ at most in the last m edges. Therefore $\mathbf{E}\left[\Delta_t\right] \le e^{O(1)}m\left(\frac{t}{\tau}\right)^{2/\alpha}$.   □

To apply Azuma's inequality we note first that

$$\left|\mathbf{E}_{G_t}[T(u)] - \mathbf{E}_{\hat{G}_t}[T(u)]\right| = \left|\mathbf{E}\left[\sum_{\rho=\tau}^{t}\sum_{i=1}^{m}(F(|u-u_i^\rho|) - F(|u-\hat{u}_i^\rho|))\right]\right| \le \mathbf{E}\left[\Delta_t(u)\right],$$

(21)

and from Lemma 3

$$\sum_{\tau=1}^{t}\mathbf{E}\left[\Delta_t(u)\right]^2 \le (e^{O(1)}mI)^2 t^{4/\alpha}\sum_{\tau=1}^{t}\tau^{-4/\alpha} = O\left(I^2 m^2(t\ln t + t^{4/\alpha})\right)$$

Therefore, there is C_1 such that

$$\mathbf{Pr}\left[|T(u) - \mathbf{E}\left[T(u)\right]| \ge C_1 Im(t^{2/\alpha} + t^{1/2}\ln t)(\ln n)^{1/2}\right] \le e^{-2\ln n} = n^{-2}.$$ □

3.3   Concentration of $d_k(t)$

We follow the proof of Lemma 3, replacing $T(u)$ by $d_k(t)$ and using the same coupling, When we reach (21) we find that $\left|\mathbf{E}_{G_t}[d_k(t)] - \mathbf{E}_{\hat{G}_t}[d_k(t)]\right| \le 2\mathbf{E}[\Delta_t]$, the rest is the same.

This proves (1) and completes the proof of Theorem 1(a).

4   Small Separators

In this section we prove Theorem 1(b). For this, we assume $\alpha > 0$ and $m_0 \le m$ where m_0 is a sufficiently large constant and $\varphi, \eta = o(1)$ are such that $\eta n \to \infty$ and $J_\eta \le \varphi I$.

We use the geometry of the instance to obtain a sparse cut. Consider partitioning the vertices in V_n using a great circle of S. This will divide V_n into sets T and $\bar{T}$ which each contain about $n/2$ vertices. More precisely, we have

$$\mathbf{Pr}\left[|T| < (1-\xi)n/2\right] = \mathbf{Pr}\left[|\bar{T}| < (1-\xi)n/2\right] \le e^{-\xi^2 n/5}.$$

To bound $e(T, \bar{T})$, the number of edges crossing the cut, we divide the edges in two types. We call an edge $\{u, v\}$ in G_n long if $|u-v| \ge \eta$, otherwise we call it short. We will show that **whp** the number of long edges is small, and therefore we just need to consider short edges in a cut. Let Z denote the number of long edges. Then

$$\mathbf{E}\left[Z\right] \leq mt_0 + m \sum_{t \geq t_0} \sum_{v \in V_t} \frac{\deg_t(v) J_\eta}{\alpha m I t} \leq mt_0 + m \sum_{t \geq t_0} \frac{J_\eta}{\alpha I} = mt_0 + O(n\varphi). \qquad \square$$

Now **whp** there are at most $\mathbf{E}\left[Z\right]/\varphi^{1/2}$ long edges. Apart from these, edges only appear between vertices within distance η, so only edges incident with vertices appearing in the strip within distance η of the great circle can appear in the cut. Since $\eta = o(1)$, this strip has area less than $3\eta\sqrt{\pi}$, and, letting U denote the vertices appearing in this strip, we have

$$\mathbf{Pr}\left[|U| \geq 4\sqrt{\pi}\eta n\right] \leq e^{-2\sqrt{\pi}\eta n/27} = o(1).$$

Even if every one of the vertices chooses its m neighbors on the opposite side of the cut, this will yield at most $4\sqrt{\pi}\eta nm$ edges **whp**. So the graph has a cut with

$$e(T, \bar{T}) = \tilde{O}((\eta + \varphi^{1/2})n)$$

with probability at least $1 - o(1)$. $\qquad \square$

The proofs of Theorems 2 and 3 are left to the full version.

Acknowledgement. We thank Henri van den Esker for pointing out some errors and for some very useful comments.

References

1. Aiello, W., Chung, F.R.K., Lu, L.: A random graph model for massive graphs. In: Proc. of the 32nd Annual ACM Symposium on the Theory of Computing, pp. 171–180 (2000)
2. Aiello, W., Chung, F.R.K., Lu, L.: Random Evolution in Massive Graphs. In: Proc. of IEEE Symposium on Foundations of Computer Science, pp. 510–519 (2001)
3. Albert, R., Barabási, A., Jeong, H.: Diameter of the world wide web. Nature 401, 103–131 (1999)
4. Barabasi, A., Albert, R.: Emergence of scaling in random networks. Science 286, 509–512 (1999)
5. Berger, N., Bollobas, B., Borgs, C., Chayes, J., Riordan, O.: Degree distribution of the FKP network model. In: Proc. of the 30th International Colloquium of Automata, Languages and Programming, pp. 725–738 (2003)
6. Berger, N., Borgs, C., Chayes, J., D'Souza, R., Kleinberg, R.D.: Competition-induced preferential attachment. In: Díaz, J., Karhumäki, J., Lepistö, A., Sannella, D. (eds.) ICALP 2004. LNCS, vol. 3142, pp. 208–221. Springer, Heidelberg (2004)
7. Blandford, D., Blelloch, G.E., Kash, I.: Compact Representations of Separable Graphs. In: Proc. of ACM/SIAM Symposium on Discrete Algorithms, pp. 679–688 (2003)
8. Bollobás, B., Riordan, O.: Mathematical Results on Scale-free Random Graphs. In: Handbook of Graphs and Networks, Wiley-VCH, Berlin (2002)

9. Bollobás, B., Riordan, O.: The diameter of a scale-free random graph. Combinatorica 4, 5–34 (2004)

10. Bollobás, B., Riordan, O.: Coupling scale free and classical random graphs. Internet Mathematics 1(2), 215–225 (2004)

11. Bollobás, B., Riordan, O., Spencer, J., Tusanády, G.: The degree sequence of a scale-free random graph process. Random Structures and Algorithms 18, 279–290 (2001)

12. Broder, A., Kumar, R., Maghoul, F., Raghavan, P., Rajagopalan, S., Stata, R., Tomkins, A., Wiener, J.: Graph structure in the web. In: Proc. of the 9th Intl. World Wide Web Conference, pp. 309–320 (2002)

13. Buckley, G., Osthus, D.: Popularity based random graph models leading to a scale-free degree distribution. Discrete Mathematics 282, 53–68 (2004)

14. Chung, F.R.K., Lu, L., Vu, V.: Eigenvalues of random power law graphs. Annals of Combinatorics 7, 21–33 (2003)

15. Chung, F.R.K., Lu, L., Vu, V.: The spectra of random graphs with expected degrees. Proceedings of national Academy of Sciences 100, 6313–6318 (2003)

16. Cooper, C., Frieze, A.M.: A General Model of Undirected Web Graphs. Random Structures and Algorithms 22, 311–335 (2003)

17. Cooper, C., Frieze, A.M., Vera, J.: Random deletions in a scale free random graph process. Internet Mathematics 1, 463–483 (2004)

18. Drinea, E., Enachescu, M., Mitzenmacher, M.: Variations on Random Graph Models for the Web, Harvard Technical Report TR-06-01 (2001)

19. Estrada, E.: Spectral scaling and good expansion properties in complex networks. Europhysics Letters 73(4), 649–655 (2006)

20. Erdös, P., Rényi, A.: On random graphs I. Publicationes Mathematicae Debrecen 6, 290–297 (1959)

21. Fabrikant, A., Koutsoupias, E., Papadimitriou, C.H.: Heuristically Optimized Trade-Offs: A New Paradigm for Power Laws in the Internet. In: Proc. of 29th International Colloquium of Automata, Languages and Programming (2002)

22. Faloutsos, M., Faloutsos, P., Faloutsos, C.: On Power-law Relationships of the Internet Topology. ACM SIGCOMM Computer Communication Review 29, 251–262 (1999)

23. Flaxman, A.: Expansion and lack thereof in randomly perturbed graphs. In: Proc. of the Web Algorithms Workshop (to appear, 2006)

24. Flaxman, A., Frieze, A.M., Vera, J.: A Geometric Preferential Attachment Model of Networks, Internet Mathematics (to appear)

25. Gómez-Gardeñes, J., Moreno, Y.: Local versus global knowledge in the Barabási-Albert scale-free network model. Physical Review E 69, 037103 (2004)

26. Hayes, B.: Graph theory in practice: Part II. American Scientist 88, 104–109 (2000)

27. Kleinberg, J.M., Kumar, R., Raghavan, P., Rajagopalan, S., Tomkins, A.S.: The Web as a Graph: Measurements, Models and Methods. In: Asano, T., Imai, H., Lee, D.T., Nakano, S.-i., Tokuyama, T. (eds.) COCOON 1999. LNCS, vol. 1627, Springer, Heidelberg (1999)

28. Kumar, R., Raghavan, P., Rajagopalan, S., Sivakumar, D., Tomkins, A., Upfal, E.: Stochastic Models for the Web Graph. In: Proc. IEEE Symposium on Foundations of Computer Science, p. 57 (2000)

29. Kumar, R., Raghavan, P., Rajagopalan, S., Sivakumar, D., Tomkins, A., Upfal, E.: The Web as a Graph. In: PODS 2000. Proc. 19th ACM SIGACT-SIGMOD-AIGART Symp. Principles of Database Systems, pp. 1–10 (2000)

30. Li, L., Alderson, D., Doyle, J.C., Willinger, W.: Towards a Theory of Scale-Free Graphs: Definition, Properties, and Implications. Internet Mathematics 2(4), 431–523

31. Mihail, M.: private communication

32. Kumar, R., Raghavan, P., Rajagopalan, S., Tomkins, A.: Trawling the Web for emerging cyber-communities. Computer Networks 31, 1481–1493 (1999)

33. McDiarmid, C.J.H.: Concentration. Probabilistic methods in algorithmic discrete mathematics, 195–248 (1998)

34. Mihail, M., Papadimitriou, C.H.: On the Eigenvalue Power Law. In: Proc. of the 6th International Workshop on Randomization and Approximation Techniques, pp. 254–262 (2002)

35. Mihail, M., Papadimitriou, C.H., Saberi, A.: On Certain Connectivity Properties of the Internet Topology. In: Proc. IEEE Symposium on Foundations of Computer Science, p. 28 (2003)

36. Mitzenmacher, M.: A brief history of generative models for power law and lognormal distributions. Internet Mathematics 1(2), 226–251 (2004)

37. Newman, M., Barabási, A.-L., Watts, D.J.: The Structure and Dynamics of Networks, Princeton University Press (2006)

38. Penrose, M.D.: Random Geometric Graphs. Oxford University Press, Oxford (2003)

39. Simon, H.A.: On a class of skew distribution functions. Biometrika 42, 425–440 (1955)

40. van der Hofstad, R.: Random Graphs and Complex Networks, unpublished manuscript (2007)

41. Watts, D.J.: Small Worlds: The Dynamics of Networks between Order and Randomness. Princeton University Press, Princeton (1999)

42. Yule, G.: A mathematical theory of evolution based on the conclusions of Dr. J.C. Willis. Philosophical Transactions of the Royal Society of London (Series B) 213, 21–87 (1925)

Clustering Social Networks

Nina Mishra[1,4], Robert Schreiber[2], Isabelle Stanton[1,*], and Robert E. Tarjan[2,3]

[1] Department of Computer Science, University of Virginia
{nmishra,istanton}@cs.virginia.edu
[2] HP Labs
{rob.schreiber,robert.tarjan}@hp.com
[3] Department of Computer Science, Princeton University
[4] Search Labs, Microsoft Research

Abstract. Social networks are ubiquitous. The discovery of close-knit clusters in these networks is of fundamental and practical interest. Existing clustering criteria are limited in that clusters typically do not overlap, all vertices are clustered and/or external sparsity is ignored. We introduce a new criterion that overcomes these limitations by combining internal density with external sparsity in a natural way. An algorithm is given for provably finding the clusters, provided there is a sufficiently large gap between internal density and external sparsity. Experiments on real social networks illustrate the effectiveness of the algorithm.

1 Introduction

Social networks have gained popularity recently with the advent of sites such as MySpace, Friendster, Facebook, etc. The number of users participating in these networks is large, e.g., a hundred million in MySpace, and growing. These networks are a rich source of data as users populate their sites with personal information. Of particular interest in this paper is the graph structure induced by the friendship links.

A fundamental problem related to these networks is the discovery of clusters or communities. Intuitively, a cluster is a collection of individuals with dense friendship patterns internally and sparse friendships externally. We give a precise definition of a cluster shortly. There are many reasons to seek tightly-knit communities in networks, for instance, target marketing schemes can be designed based on clusters, and it has been claimed that terrorist cells can be identified [12].

What is a good cluster in a social network? There are numerous existing criteria for defining good graph clusters, and accompanying each criterion is a multitude of algorithms. One popular criterion is based on finding clusters of high conductance. The conductance of a cut A, B is the ratio of the number of edges crossing the cut to the minimum of the volume of A and B, where the volume of A is the number of edges emanating from the vertices in A. The conductance

---

* Supported by an NPSC Graduate Fellowship and Google Anita Borg Scholarship.

A. Bonato and F.R.K. Chung (Eds.): WAW 2007, LNCS 4863, pp. 56–67, 2007.

of a cluster is the minimum conductance of any cut in the cluster. A spectral algorithm is typically used to discover these clusters where the eigenvector of a matrix related to the adjacency matrix can be used to find a good cut of the graph into subgraphs A, B. The process is then recursively repeated (on A and B) until k clusters are found (where k is an input parameter) or until the conductance of the next best cut is larger than some threshold. Formal guarantees can be proved for some variants of this basic algorithm [9].

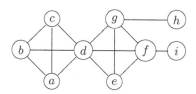

Fig. 1. Overlapping clusters

Cut-based graph clustering algorithms produce a strict partition of the graph. This is particularly problematic for social networks as illustrated in Fig. 1. In this graph, d belongs to two clusters $\{a, b, c, d\}$ and $\{d, e, f, g\}$. Furthermore, h and i need not be clustered. A cut-based approach will either put $\{a, b, c, d, e, f, g\}$ into one cluster, which is not desirable since e, f, g have no edges to a, b, c, or cut at d, putting d into one of the clusters, say $\{a, b, c, d\}$, but leaving d out of $\{e, f, g\}$ which then leaves a highly connected vertex outside of the cluster.

This example motivates a new formulation of the graph clustering problem that does not stipulate that each vertex belong to exactly one cluster. Our objective is to identify clusters that are internally dense, i.e., each vertex in the cluster is adjacent to at least a β-fraction of the cluster, and externally sparse, i.e., any vertex outside of the cluster is adjacent to at most an α-fraction of the vertices in the cluster. For a vertex v and a subset of vertices C, the notation $E(v, C)$ below denotes the set of edges between v and C.

Definition 1. *Given a graph, $G = (V, E)$, where every vertex has a self-loop[1] $C \subset V$ is an (α, β)-cluster if*

1. **Internally Dense:** $\forall v \in C, |E(v, C)| \geq \beta|C|$
2. **Externally Sparse:** $\forall u \in V \setminus C, |E(u, C)| \leq \alpha|C|$

Given $0 \leq \alpha < \beta \leq 1$, the (α, β)-clustering problem is to find all (α, β)-clusters.

The new clustering criterion does not seek a strict partitioning of the data. To see why clusters can overlap, return to Fig. 1. Both $\{a, b, c, d\}$ and $\{d, e, f, g\}$ are $(\frac{1}{4}, 1)$-clusters. Furthermore, h and i do not fall into an (α, β)-cluster if $0 \leq \alpha < \frac{1}{2} < \beta \leq 1$, and consequently would not be clustered.

---

[1] This is a technical assumption needed to ensure that $\beta = 1$ clusters are possible.

Observe that as $\beta \to 1$, the cluster C approaches a clique and as $\alpha \to 0$, C tends to a disconnected component. We want $\alpha < \beta$, since we want vertices outside of a cluster to have fewer neighbors in the cluster than vertices that belong to the cluster.

While we use social networks as a motivating context, our problem statement and algorithms apply to the more general context of graph clustering.

Contributions. We begin by investigating combinatorial properties of (α, β)-clusters. We bound the extent to which two clusters can overlap. For two clusters of equal size, we show that they overlap in at most $\min\{1-(\beta-\alpha), \alpha/(2\beta-1)\}|C|$ vertices. For certain values of α and β, it is possible for one cluster to be contained in another. However, we show that if the ratio of the size of the largest cluster to the smallest cluster is at most $\frac{1-\alpha}{1-\beta}$ then one cluster cannot be contained in another. Finally, we give a loose upper bound on the number of $(\alpha, 1)$-clusters of size s, $O((n/s)^{\alpha s+1})$, where n is the number of vertices.

Next, we introduce the notion of a ρ-champion of a cluster: a vertex with at most $\rho|C|$ neighbors outside of the cluster C. We prove that in the case that there is a large gap between $\alpha/2$ and β i.e., $\beta > \frac{1}{2} + \frac{\rho+\alpha}{2}$, there can be at most n (α, β)-clusters with ρ-champions of a given cluster size and there is a simple deterministic algorithm for finding all such clusters in time $O(m^{0.7}n^{1.2}+n^{2+o(1)})$, where m is the number of edges.

To determine whether the theoretical constructs we introduced actually exist in practice, we tested our algorithms on three real networks: High Energy Physics Co-authors, Theory Co-authors and Live Journal. Experiments show that our algorithm is able to find 90% of the ground-truth clusters of practical interest more quickly than previous algorithms.

2   Related Work

Our (α, β)-clustering formulation is new, but has been considered in restricted settings under different guises. The problem of finding the $(0, \beta)$-clusters in a graph can be reduced to first finding connected components and then outputting the components that are β-connected. This problem can be solved efficiently via depth first search in $O(|E| + |V|)$ time for a graph $G = (V, E)$. Also, the problem of finding $(1 - \frac{1}{n}, 1)$-clusters is equivalent to finding the maximal cliques in a graph. This problem has a rich history. Known algorithms find all maximal cliques in time that depends polynomially on the size of the graph and the number of maximal cliques [18,8].

The problem of finding $((1 - \epsilon)\beta, \beta)$-clusters, for small ϵ, has also been studied under the name of finding quasi-cliques. Abello et al. [1] present a method for finding subgraphs with average connectivity β. Hartuv and Shamir [6] find densely connected subgraphs where $\beta > 1/2$ via a min-cut algorithm. These algorithms do not consider an external sparsity (α) criterion. We will give an example (Fig. 2) where if these algorithms were used to find $(1/n, 1-1/2n)$-clusters (of which there is only 1), they return 2^n ($\frac{n-1}{n}, 1$)-clusters.

Spectral clustering is a very popular method that involves recursively splitting the graph using various criteria, e.g., the principal eigenvector of the adjacency matrix. Successful approaches have been employed by [9,16,10,17,15], among many others. All of these approaches do not allow overlapping clusters which is one of the main goals of our work.

Newman and others have advocated modularity as an optimization criterion for graph partitioning [15]. The modularity of a partition is the amount by which the number of edges between vertices in the same subset exceeds the number predicted by the degree-distribution preserving random graph model of Chung [2]. Newman proposed several methods for optimizing modularity, among them a spectral approach, and others have found competitive methods as well.

Flake et al. [4] use a recursive cut approach intended to optimize the expansion of the clustering but use Gomory-Hu trees [5] to find the cut instead of eigenvectors. The expansion of a cut is very similar to the conductance of a cut. The minimum quality of the clustering is guaranteed by adding a sink to the graph. Again, the goal of this work is different from ours in that a partitioning is constructed, disallowing overlapping clusters.

Modeling flow through a network is another way to cluster a graph [4,3]. MCL models flow through two alternating Markov processes, expansion and inflation. MCL has been widely used for clustering in biological networks but requires that the graph be sparse and only finds overlapping clusters in restricted cases. (α, β)-Clustering has no restrictions on the general structure of the graph and allows clusters of different sizes to overlap.

There has also been considerable work in finding communities on the web. Kumar et al. [13] approach the problem as one of finding bicliques as the cores of communities. While our approach can be adapted to find bicliques, we deal with more general community structures.

3   Combinatorics of (α, β)-Clusters

In this section, we discuss various combinatorial properties of (α, β)-clusters including cluster overlap, containment and number of clusters.

Prior to doing so, we make a quick remark about the value of β. If $\beta < \frac{1}{2}$ then it is possible to have a cluster containing two disconnected components, i.e., a subset of vertices with a cut of size 0 could form a cluster. Imagine two cliques K_n with no edges in between them. If $\beta < \frac{1}{2}$ then these two disconnected cliques form one $(0, \frac{1}{2})$-cluster. Consequently, we insist that $\beta > \frac{1}{2}$. In that case, a cluster is necessarily connected; select any two vertices u and u' in the cluster, since u is adjacent to more than half of the cluster and so is u', there must be at least one vertex that they both neighbor. Thus, there is a path of length at most two between any two vertices in a cluster. We will use this fact later in some of our analysis. In this paper, we assume that $\beta > 1/2$ so that all clusters are connected, although it would be interesting to consider other restrictions that enforce connectedness.

Notation. We use the following notation to describe our results. For a graph $G = (V, E)$, n denotes the number of vertices and m denotes the number of edges. For a subset of vertices $A \subseteq V$, $|A|$ denotes the number of vertices in A. $E(v, A)$ denotes the set of edges between a vertex v and a subset of vertices A. The neighbors of a vertex v are denoted by $\Gamma(v)$. The function $\tau(v) = \Gamma(v) \cup \Gamma(\Gamma(v))$ indicates all neighbors of path distance 1 or 2 from v.

Cluster Overlap. Given two (α, β)-clusters A, B where $|A| \geq |B|$, we now determine the maximum size of the overlap, namely $|A \cap B|$. In the case where $\beta = 1$, $|A \cap B|$ can be no larger than $\alpha|B|$ (otherwise, there would be a vertex outside of B that is adjacent to more than α of B). Alternatively, in the case where $\alpha = 0$, $|A \cap B|$ must be 0. More generally, we seek a bound for arbitrary values of α and β. We express the overlap as the fraction of vertices in A, i.e., $\gamma = \frac{|A \cap B|}{|A|}$.

Proposition 1. *For two (α, β)-clusters, A and B, where $|A| \geq |B|$, an upper bound on the ratio of the intersection, $|A \cap B|$, to the larger one, $|A|$, is $\gamma = \min(1 - (\beta - \alpha\frac{|B|}{|A|}), \frac{\alpha}{2\beta-1}\frac{|B|}{|A|})$. When $\beta - \alpha\frac{|B|}{|A|} > \frac{1}{2}$, $\frac{\alpha}{2\beta-1}\frac{|B|}{|A|}$ is the minimum and otherwise $1 - (\beta - \alpha\frac{|B|}{|A|})$ is the minimum.*

Cluster Containment. Given that clusters can overlap, it is natural to ask if one cluster can be contained in another. In some circumstances, α and β may be such that clusters are contained in each other. For example, consider two cliques, C and D, each containing n vertices. Assume that each vertex in C is adjacent to two vertices in D. When $\beta = \frac{1}{2} + \frac{2}{n}$ and $\alpha = \frac{2}{n}$, $C \cup D$ is an (α, β) cluster that contains both C and D.

If we want to prevent our algorithm from finding clusters where one is contained in another, we can do so by requiring that the ratio of the largest to the smallest cluster is at most $\frac{1-\alpha}{1-\beta}$.

Corollary 1. *Let A and B be (α, β)-clusters and assume that $|B| \leq |A|$. If $\frac{|A|}{|B|} < \frac{1-\alpha}{1-\beta}$ then B can not be contained in A.*

The larger the gap between α and β, the larger the bound. For example, if $\alpha = 1/4$ and $\beta = 3/4$, then the larger cluster must be at least 3 times larger than the smaller before the smaller can be contained in the larger. Similarly, if $\alpha = 1/8$ and $\beta = 7/8$ then the ratio is 7.

Bounding the Number of $(\alpha, 1)$-clusters. We next consider the problem of upper bounding the number of $(\alpha, 1)$-clusters. We give a superpolynomial bound on the number of clusters of a fixed size $s = f(n)$. More generally, it would be interesting to bound the number of possible (α, β)-clusters, but our analysis here is focused on cliques.

We wish to bound the number of $(\alpha, 1)$-clusters of size $s = f(n)$ in a graph $G = (V, E)$ where $|V| = n$. We know that no two clusters can overlap in more than αs vertices from Prop. 1.

Proposition 2. *Let $G = (V, E)$ where $|V| = n$. If $\mathcal{C}$ is the set of $(\alpha, 1)$-clusters of size s in G then $|\mathcal{C}| = O((\frac{n}{s})^{\alpha s + 1})$.*

Proof. From Prop. 1, two clusters of size s can share at most αs vertices. Let us ignore the edges in the graph and consider clusters as subsets of vertices. Now we can say that every subset of size $\alpha s + 1$ must appear in at most one set in our collection. There are a total of $\binom{n}{s}$ subsets of size s and each of these subsets contains $\binom{s}{\alpha s + 1}$ subsets of size $\alpha s + 1$. By simple combinatorics we can have at most $\binom{n}{\alpha s + 1}/\binom{s}{\alpha s + 1}$ clusters of size s. The bound $|\mathcal{C}| \leq \binom{n}{\alpha s + 1}/\binom{s}{\alpha s + 1} = O((\frac{n}{s})^{\alpha s + 1})$ follows from Stirling's Approximation.[2] $\square$

We note that this bound is tight when $\alpha = 0$ and when α approaches 1. If we let $\alpha = 0$ then the bound indicates that the number of clusters is at most $\frac{n}{k}$. This is tight because clusters cannot overlap at all. At the other extreme, consider the complement of the graph shown in Fig. 2. Let $\alpha = \frac{n-1}{n}$ and $\beta = 1$. For $k = n$ the bound on the number of clusters from our bound is 2^n. This number is realized since the set of clusters is $B = \{b_1 \ldots b_n | b_i = x_i \vee y_i\}$. $|B| = 2^n$ so the bound is tight in this case. We can construct a graph of this type for all α of the form $\frac{n-1}{n}$ so we have a tight exponential bound for these α values.

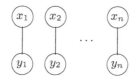

Fig. 2. A graph G where $\overline{G}$ has exponentially many clusters

We believe that the bound given in Prop. 2 overcounts the number of clusters when $\alpha \leq \frac{1}{2}$. We note that our examples of graphs that meet the exponential bound all have $\alpha \geq \frac{1}{2}$. Consider the case where we have two $(\alpha, 1)$-clusters, A and B that overlap in αs vertices. Let D be a third cluster such that $|A \cap D| = |B \cap D| = \alpha s$ but $A \cap B \cap D = \emptyset$. This is allowed by the construction in Prop. 2. Let $u \in A \cap C$ and $v \in B \cap C$. Since $u, v \in C$ and $\beta = 1$ $(u, v) \in E$. However, u is already connected to $\alpha |B|$ in the form of $A \cap B$, so we have an α violation. Therefore, we counted D as an (α, β)-cluster when we should not have.

Another criticism of counting $(\alpha, 1)$-clusters with Prop. 2 is that edges are completely ignored. Consider K_4 where $s = 3$ and $\alpha = 1/3$. The bound allows 3 clusters of size 3. In reality, due to α violations, there are none.

4   An Algorithm for Finding Clusters with Champions

In this section, we make some restrictions to the general (α, β)-clustering problem, motivate these restrictions and then give an algorithm for finding clusters of

---

[2] This exactly corresponds to the construction of a Steiner System.

this restricted form. Specifically, we first justify a gap between internal density and external sparsity. Next, we introduce the notion of a champion of a cluster. Intuitively, a vertex champions a cluster if it has more affinity into the cluster than out of it. We then give a simple, deterministic algorithm for finding all (α, β)-clusters with ρ-champions in a graph, assuming that $\beta > \frac{1}{2} + \frac{\rho+\alpha}{2}$.

Gap Between Internal Density and External Sparsity. To motivate a gap between internal density and external sparsity, consider Fig. 2. Observe that depending on the choice of α and β, the number of clusters may be exponential in the size of the graph. In practice, an algorithm that outputs more clusters than vertices is quite undesirable – especially given that social networks are massively large data sets. Thus, we seek a restriction that will reduce the number of clusters. The restriction considered in this paper is a large gap between β and $\alpha/2$.

Champions. To motivate champions, observe that for $\overline{G}$ of G given in Fig. 2, each vertex in each cluster has as many neighbors outside the cluster as within it. There is no vertex that "champions" the cluster in the sense that many of its neighbors are in the cluster. For example, theoretical physicists form a community in part because there are some champions that have more friends that are theoretical physicists than not. Specifically, if *every* vertex in a subset A has as many neighbors out of A as into A, then it is arguable if A is really even a cluster. This motivates us to formally define the notion of a ρ-champion.

Definition 2. *A vertex $c \in C$ ρ-champions a cluster C if $|\Gamma(c) \cap V \setminus C| \leq \rho|C|$, for some $0 \leq \rho \leq 1$.*

Deterministic Algorithm. We now claim that if $\beta > \frac{1}{2} + \frac{\rho+\alpha}{2}$ or $\alpha < (2\beta-1)(\beta-\rho)$ then there are at most n clusters with ρ-champions and further that there is a simple deterministic algorithm for finding the clusters. In the following, we make the simplifying assumption that every cluster has the same size. The lemma can be suitably modified in the case of clusters of different sizes.

Lemma 1. *If either $\beta > \frac{1}{2} + \frac{\rho+\alpha}{2}$ or $\alpha < (2\beta-1)(\beta-\rho)$ then there are at most n (α, β)-clusters of size s with ρ-champions.*

A large gap between β and $\frac{1}{2} + \frac{\alpha+\rho}{2}$ yields a simple algorithm for deterministically pinning down all the clusters. Let the input to the algorithm be α, β, the graph G and the size s of the clusters to be found.

Algorithm 1. Deterministic Clustering Algorithm, when $\beta > \frac{1}{2} + \frac{\alpha+\rho}{2}$.

1: Input: α, β, s, G
2: **for** each $c \in V$ **do**
3:     $C = \emptyset$
4:     **for** each $v \in \tau(c)$ **do**
5:         **if** $|\Gamma(v) \cap \Gamma(c)| \geq (2\beta-1)s$ then add v to C.
6:     **end for**
7:     **if** C is an (α, β)-cluster then output C.
8: **end for**

The following lemma shows that if v and c share sufficiently many neighbors, then v is necessarily part of the cluster C that c champions.

Lemma 2. *Let C be an (α, β)-cluster and c its ρ-champion. Let $\beta > \frac{1}{2} + \frac{\rho + \alpha}{2}$. A vertex v is in the cluster C if and only if $|\Gamma(v) \cap \Gamma(c)| \geq (2\beta - 1)|C|$.*

When the size of the cluster is fixed, Lemma 2 also implies that C is unique. Since we can bound the number of clusters of each size to n, we can also bound the total number of (α, β)-clusters with ρ-champions to be $O(n^2)$. Additional bounds to guarantee uniqueness when the size of the cluster is allowed to vary can be easily obtained.

Consequently, we have the following theorem.

Theorem 1. *Let $G = (V, E)$ be a graph and $\beta > \frac{1}{2} + \frac{\rho + \alpha}{2}$. Algorithm 1 exactly finds all (α, β)-clusters of size s that have ρ-champions in time $O(m^{0.7}n^{1.2} + n^{2+o(1)})$.*

To interpret the theorem, when clusters have ρ-champions where $\rho = \alpha$, a separation of $\frac{1}{2}$ is needed between β and α in order for the algorithm to find all the clusters. The worse the champion, the fewer the number of valid α and β values where the algorithm is guaranteed to succeed. For example, if $\rho = 3\alpha$ then the gap between β and α must be larger, namely $\beta > 2\alpha + \frac{1}{2}$.

The running time follows from the fact that the algorithm computes the number of neighbors that each pair of vertices share. We can precompute $|\Gamma(v_i) \cap \Gamma(v_j)|$ for all $i, j \in V$ by noting that if A is the adjacency matrix of G then $(A^T A)_{i,j} = |\Gamma(v_i) \cap \Gamma(v_j)|$. Yuster and Zwick [19] show that matrix multiplication can be performed in $O(m^{0.7}n^{1.2} + n^{2+o(1)})$ time. Checking the α, β conditions requires $O(m^{0.7}n^{1.2} + n^{2+o(1)} + n(\tau(c) + n)) = O(m^{0.7}n^{1.2} + n^{2+o(1)})$ time.

In the case G is a typical social network, G has small average degree and A is a sparse matrix . If we let d be the average degree of the graph then $m = dn/2$. Thus, for small d, the algorithm runs in $O(d^{0.7}n^{1.9} + n^{2+o(1)})$ time.

5 Experiments

We introduced the notion of a ρ-champion and gave an algorithm for finding (α, β)-clusters with ρ-champions. A natural next question is: Do (α, β)-clusters with ρ-champions even exist in real graphs? And, if so, do most (α, β)-clusters have ρ-champions? To answer the first question, we study three real networks induced by co-authorship among high energy physicists, co-authorship among theoretical computer scientists, as well as a real, online social networking site known as LiveJournal. To answer the second question, we need an algorithm that can find (α, β)-clusters independent of whether they have ρ-champions. The best previous algorithm for this problem is due to Tsukiyama et al [18] that finds all maximal cliques in a graph, i.e., all $(\alpha, 1)$-clusters.

Our experiments uncovered a few surprising facts. First, our simple algorithm was able to find $\approx 90\%$ of the maximal cliques in these graphs where $\alpha \leq \frac{1}{2}$. Next, among the cliques we missed, we found that there was no strong ρ-champion. Finally, our algorithm was orders of magnitudes faster than Tsukiyama's. In short, our algorithm more quickly discovers clusters of practical interest, i.e., small α, small ρ and large β.

Data Sets and Tsukiyama's Algorithm. As mentioned, three data sets were used: the High Energy Physics Theory Co-Author graph (HEP) [7], the Theory Co-Author graph (TA) and a subset of the LiveJournal graph (LJ) [14]. LiveJournal is a website that allows users to create weblogs and befriend other LiveJournal users. We obtained a crawl of a subset of this site. In our graph the vertices correspond to usernames and the edges to friendships. In the Theory and HEP Co-Author graphs, authors are vertices and edges correspond to co-authors. Some basic statistics about these graphs are given below.

Data set	Size	Avg Deg.	Min Deg.	Max Deg.	Avg $\tau(v)$	Min $\tau(v)$	Max $\tau(v)$
HEP	8,392	4.86	1	63	40.58	2	647
TA	31,862	5.75	1	567	172.85	1	8,116
LJ	581,220	11.68	1	1,980	206.15	6	15,525

Tsukiyama's algorithm finds all maximal cliques in a graph via an inductive characterization: given the maximal cliques involving the first i vertices, the algorithm shows how to extend this set to the maximal cliques involving the first $i + 1$ vertices. The algorithm's running time is polynomial in the size of the graph and the number of maximal cliques. More details can be found in [18].

Results. In this section we present numerical results comparing the ground truth of Tsukiyama's Algorithm with our Algorithm 1. For this experiment we were only interested in cliques of size 5 or larger with α values of 0.5 or less. These are the cliques that Algorithm 1 could reasonably find. We pruned the output of Tsukiyama's algorithm to contain just these cliques. We found that the HEP graph had a total of 126 cliques satisfying this definition; our algorithm found 115, or 91%. Similarly, the Theory graph had 854 cliques and our algorithm found 797 or 93%. In Figures 3, 4 and 5 we show the α and ρ distributions of the cliques found by Tsukiyama compared with the α distribution of those found by Algorithm 1. When a bar is cut off a number is placed next to the bar to indicate the true value. Bars have only been cut off when Algorithm 1 found all of the cliques that Tsukiyama's Algorithm found.

In both Theory and HEP, the distribution of ρ-values among the clusters found is exactly as our theorems predict, i.e., ρ is almost always less than $\frac{1}{2}$. And, interestingly, for LiveJournal, the distribution of ρ-values is better than our theorems predict in that we find 876 clusters where ρ is larger than $1/2$. Indeed, we find some clusters where ρ is as large as 1.2.

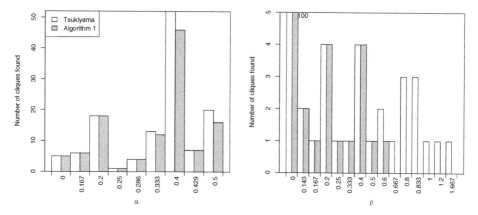

Fig. 3. For the HEP graph, α and ρ distributions are shown for the cliques found by Tsukiyama's algorithm vs. the cliques found by Algorithm 1. Our algorithm found 115 out of 126 maximal cliques.

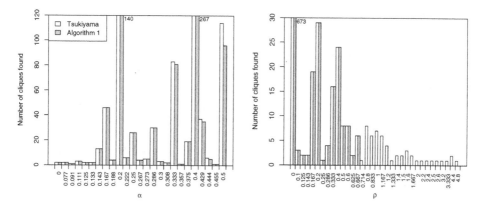

Fig. 4. For the TA graph, α and ρ distributions are shown for the cliques found by Tsukiyama's Algorithm vs. the cliques found by Algorithm 1. Our algorithm found 797 out of 854 maximal cliques.

Timing. Our experiments were run on a machine with 2 dual core 3 GHz Intel Xeons and 16 Gigabytes of RAM. We report wall-clock time for all of our experiments.

Experiment	HEP	TA	LJ
Alg. 1, $(\alpha, \beta) = (0.5, 1)$	8 secs	2 min 4 sec	3 hours 37 min
Tsukiyama	8 hours	36 hours	N/A

Note that after one week of running Tsukiyama et al.'s algorithm on the LJ data set, the algorithm did not complete. In fact, only 6% of the graph had been considered. However, our Algorithm 1 found 4289 cliques of size greater than 5 with $\alpha \leq .5$ in a few hours.

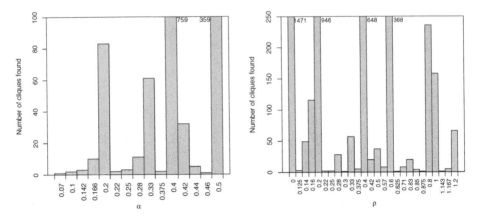

Fig. 5. α and ρ distributions for Algorithm 1 for the LJ graph. Tsukiyama's algorithm was too slow to generate ground truth results for this graph. Our Algorithm 1 found 4289 cliques.

6   Summary and Future Work

We introduced a new criterion for discovering overlapping clusters that captures intuitive notions of internal density and external sparsity. We also give a deterministic algorithm for discovering clusters assuming each cluster has a champion and there is a sufficiently large gap between internal density and external sparsity. Experiments indicate that our algorithm succeeds in finding good clusters.

While we assume $\beta > \frac{1}{2}$ to enforce cluster connectedness, we believe this assumption is too strong. In particular, a subgraph can be connected while β is much less than $\frac{1}{2}$, e.g., a long cycle. Furthermore, $\beta > \frac{1}{2}$ precludes our algorithm from finding very large clusters because the average degree of a vertex in a social network is typically small.

Generalizations of (α, β)-clustering to weighted and directed graphs are also of interest. Our work assumes that edges are not weighted. But in real social networks, there is a strength of connectivity between pairs of individuals corresponding to how often they communicate. This weight could be exploited in the discovery of close-knit communities. In addition, some networks induce directed graphs. For example, the direction of edges in email networks plays an important role in definining communities otherwise spam mailers would belong to every cluster.

Decentralized and streaming algorithms are essential for modern networks such as instant messaging or email graphs. In particular, it is often difficult to even collect the graph in one centralized location [11]. Thus, algorithms that can compute clusters with only local information are needed. Further, given that social networks are dynamic data sets, i.e., vertices and edges come and go, streaming graph clustering algorithms are an important avenue for future research.

Acknowledgments. We are grateful to Mark Sandler for early discussions, Ying Xu for the LiveJournal crawl and the anonymous reviewers for their useful comments.

References

1. Abello, J., Resende, M.G.C., Sudarsky, S.: Massive quasi-clique detection. In: Rajsbaum, S. (ed.) LATIN 2002. LNCS, vol. 2286, pp. 598–612. Springer, Heidelberg (2002)
2. Aiello, W., Chung, F., Lu, L.: A random graph model for massive graphs. In: STOC 2000. Proceedings of the 32nd Annual ACM Symposium on Theory of Computing, Portland, Oregon, pp. 171–180 (May 21-23, 2000)
3. Van Dongen, S.: A new cluster algorithm for graphs. Technical report, Universiteit Utrecht (July 10, 1998)
4. Flake, G.W., Tarjan, R.E., Tsioutsiouliklis, K.: Graph clustering and minimum cut trees. Internet Mathematics 1(4), 385–408 (2004)
5. Gomory, R.E., Hu, T.C.: Multi terminal network flows. Journal of the Society for Industrial and Applied Mathematics 9, 551–571 (1961)
6. Hartuv, E., Shamir, R.: A clustering algorithm based on graph connectivity. IPL: Information Processing Letters 76, 175–181 (2000)
7. KDD Cup 2003 HEP-TH (2003), http://www.cs.cornell.edu/projects/kddcup/datasets.html
8. Johnson, D.S., Papadimitriou, C.H., Yannakakis, M.: On generating all maximal independent sets. Information Processing Letters 27(3), 119–123 (1988)
9. Kannan, R., Vempala, S., Vetta, A.: On clusterings — good, bad and spectral. In: Proceedings of the 41th Annual Symposium on Foundations of Computer Science, pp. 367–377 (2000)
10. Karypis, G., Kumar, V.: A parallel algorithm for multilevel graph partitioning and sparse matrix ordering. J. Parallel Distrib. Comput. 48(1), 71–95 (1998)
11. Kempe, D., McSherry, F.: A decentralized algorithm for spectral analysis. In: STOC-2004. Proceedings of the thirty-sixth annual ACM Symposium on Theory of Computing, pp. 561–568. ACM Press, New York (June 13-15, 2004)
12. Krebs, V.: Uncloaking terrorist networks. First Monday 7(4) (2002)
13. Kumar, R., Raghavan, P., Rajagopalan, S., Tomkins, A.: Trawling the Web for emerging cyber-communities. Computer Networks 31(11-16), 1481–1493 (1999)
14. LiveJournal, http://www.livejournal.com
15. Newman, M.E.J.: Modularity and community structure in networks. National Academy of Sciences 103, 8577–8582 (2006)
16. Shi, J., Malik, J.: Normalized cuts and image segmentation. IEEE Trans. Pattern Analysis and Machine Intelligence 22(8), 888–905 (2000)
17. Spielman, D.A., Teng, S.: Spectral partitioning works: Planar graphs and finite element meshes. In: Proceedings of the 37th Annual Symposium on Foundations of Computer Science, vol. 37, pp. 96–105 (1996)
18. Tsukiyama, S., Ide, M., Ariyoshi, H., Shirakawa, I.: A new algorithm for generating all the maximal independent sets. SIAM J. Comput. 6(3), 505–517 (1977)
19. Yuster, R., Zwick, U.: Fast sparse matrix multiplication. ACM Transactions on Algorithms 1(1), 2–13 (2005)

Manipulation-Resistant Reputations Using Hitting Time*

John Hopcroft and Daniel Sheldon

Cornell University

Abstract. Popular reputation systems for linked networks can be manipulated by spammers who strategically place links. The reputation of node v is interpreted as the world's opinion of v's importance. In PageRank [4], v's own opinion can be seen to have considerable influence on her reputation, where v expresses a high opinion of herself by participating in short directed cycles. In contrast, we show that expected hitting time — the time to reach v in a random walk — measures essentially the same quantity as PageRank, but excludes v's opinion. We make these notions precise, and show that a reputation system based on hitting time resists tampering by individuals or groups who strategically place outlinks. We also present an algorithm to efficiently compute hitting time for all nodes in a massive graph; conventional algorithms do not scale adequately.

1 Introduction

Reputation and ranking systems are an essential part of web search and e-commerce. The general idea is that the reputation of one participant is determined by the endorsements of others; for example, one web page endorses another by linking to it. However, not all participants are honorable — e.g., spammers will do their best to manipulate a search engine's rankings. A natural requirement for a reputation system is that individuals should not be able to improve their own reputation using simple self-endorsement strategies, like participating in short cycles to boost PageRank. Since PageRank enjoys many nice properties, it is instructive to see where things go wrong.

Let $G = (V, E)$ be a directed graph (e.g, the web). PageRank assigns a score $\pi(v)$ to each node v, where π is defined to be the stationary distribution of a random walk on G, giving the pleasing interpretation that the score of page v is the fraction of time a web surfer spends there if she randomly follows links forever. For technical reasons, the random walk is modified to restart in each step with probability α, jumping to a page chosen uniformly at random. This ensures that π exists and is efficient to compute. Then a well-known fact about

---

* This material is based upon work supported by the National Science Foundation under Grant No. 0514429, and by the AFOSR under Award No. FA9550-07-1-0124. Any opinions, findings and conclusions or recommendations expressed in this material are those of the author(s) and do not necessarily reflect the views of the National Science Foundation (NSF) or the AFOSR.

Markov chains [1] says that $1/\pi(v)$ is equal to the expected *return time* of v, the number of steps it takes a random walk starting at v to return to v. A heuristic argument for this equivalence is that a walk returning to v every r steps on average should spend $1/r$ of all time steps there.

Despite its popularity as a ranking system, one can easily manipulate return time by changing *only outlinks*. Intuitively, a node v should link only to nodes from which a random walk will return to v quickly (in expectation). By partnering with just one other node to form a 2-cycle with no other outlinks, v ensures a return in two steps — the minimum possible without self-loops — unless the walk jumps first. In this fashion, v can often boost its PageRank by a factor of 3 to 4 for typical settings of α [6]. However, this strategy relies on manipulating the portion of the walk before the first jump: the jump destination is independent of v's outlinks, and return time is determined once the walk reaches v again, so v's outlinks have no further effect. This suggests eliminating the initial portion of the walk and measuring reputation by the time to hit v following a restart, called the *hitting time* of node v from the restart distribution. This paper develops a reputation system based on hitting time that is provably resistant to manipulation. Our main contributions are:

- In Theorem 1, we develop a precise relationship between expected return time and expected hitting time in a random walk with restart, and show that the expected hitting time of v is equal to $(1 - p)/\alpha p$, where p is the probability that v is reached before the first restart. We will adopt p as our measure of the reputation of v.
- We prove that the resulting reputation system resists manipulation, using a natural definition of influence. For example, node v has a limited amount of influence that depends on her reputation, and she may spread that influence using outlinks to increase others' reputations. However, node v cannot alter her own reputation with outlinks, nor can she damage w's reputation by more than her original influence on w. Furthermore, the advantage that v gains by purchasing new nodes, often called *sybils* of v, is limited by the restart probability of the sybils.
- We present an efficient algorithm to simultaneously compute hitting time for all nodes. In addition to one PageRank calculation, our algorithm uses Monte Carlo sampling with running time that is linear in $|V|$ for given accuracy and confidence parameters. This is a significant improvement over traditional algorithms, which require a large-scale computation for each node.[1]

The rest of the paper is structured as follows. In section 2 we discuss related work. In section 3 we present Theorem 1, giving the characterization of hitting time that is the foundation for the following sections. In section 4 we develop a reputation system using hitting time and show that it is resistant to manipulation. In section 5 we present our algorithm for hitting time.

---

[1] Standard techniques can simultaneously compute hitting time from all possible sources to a single target node using a system of linear equations. However, what is desired for reputation systems is the hitting time from one source, or in this case a distribution, to all possible targets.

2   Related Work

Since the introduction of PageRank [4], it has been adapted to a variety of applications, including personalized web search [22], web spam detection [14], and trust systems in peer-to-peer networks [17]. Each of these uses the same general formulation and our work applies to all of them.

Much work has focused on the PageRank system itself, studying computation methods, convergence properties, stability and sensitivity, and, of course, implementation techniques. See [18] for a survey of this wide body of work. Computationally, the Monte Carlo methods in [8] and [2] are similar to our algorithms for hitting time. They use a probabilistic formulation of PageRank in terms of a *short* random walk that permits efficient sampling. In particular, we will use the same idea as [8] to efficiently implement many random walks simultaneously in a massive graph, without requiring random access.

Recent works have addressed the manipulability of PageRank: how can a group of selfish nodes place outlinks to optimize their PageRank, and how can we detect such nodes [3, 6, 11, 12, 13, 20, 23]? In particular, [3, 6, 12] all describe the manipulation strategy mentioned in the introduction.

For a more general treatment of reputation systems in the presence of strategic agents, see [9] for a nice overview with some specific results from the literature. Cheng and Friedman [5] prove an impossibility result that relates to our work — a wide class of reputation systems (including ours) cannot be resistant to a particular attack called the *sybil attack* [7]. However, their definition of resistance is very strong, requiring that no node can improve its ranking using a sybil attack; our results can be viewed as positive results under a relaxation of this requirement by limiting the damage caused by a sybil attack. We will discuss sybils in section 4.3.

Hitting time is a classical quantity of interest in Markov chains. See chapter 2 of [1] for an overview. The exact terminology and definitions vary slightly: we define hitting time as a random variable, but sometimes it is defined as the expectation of the same random variable. Also, the term *first passage time* is sometimes used synonymously. In a context similar to ours, hitting time was used as a measure of proximity between nodes to predict link formation in a social network [19]; also, the node similarity measure in [16] can be formulated in terms of hitting time.

Finally, the relationship between hitting time and return time in a random walk with restart is related to regenerative stochastic processes. In fact, Theorem 1 can be derived as a special case of a general result about such processes. See equation (15) in [15] and the references therein for details.

3   Characterizing Hitting Time

This section paves the way toward a reputation system based on hitting time by stating and proving Theorem 1. Part (i) of the theorem relates expected hitting time to expected return time — the two are essentially the same *except* for nodes where the random walk is likely to return before jumping, the sign of a known manipulation strategy. Part (ii) proves that the expected hitting time of v is completely determined by the probability that v is reached before the first jump; this will lead to precise notions of manipulation-resistance in section 4.

3.1   Preliminaries

Let $G = (V, E)$ be a directed graph. Consider the *standard random walk* on G, where the first node is chosen from starting distribution q, then at each step the walk follows an outgoing link from the current node chosen uniformly at random. Let $\{X_t\}_{t \geq 0}$ be the sequence of nodes visited by the walk. Then $\Pr[X_0 = v] = q(v)$, and $\Pr[X_t = v \mid X_{t-1} = u] = 1/\text{outdegree}(u)$ if $(u, v) \in E$, and zero otherwise. Here, we require $\text{outdegree}(u) > 0$.[2] Now, suppose the walk is modified to restart with probability α at each step, meaning the next node is chosen from the starting distribution (henceforth, *restart distribution*) instead of following a link. The new transition probabilities are:

$$\Pr[X_t = v \mid X_{t-1} = u] = \begin{cases} \alpha q(v) + \dfrac{1-\alpha}{\text{outdegree(u)}} & \text{if } (u, v) \in E \\ \alpha q(v) & \text{otherwise} \end{cases}.$$

We call this the α-*random walk* on G, and we parametrize quantities of interest by the restart probability α. A typical setting is $\alpha = 0.15$, so a jump occurs every $1/.15 \approx 7$ steps in expectation. The *hitting time* of v is $H_\alpha(v) = \min\{t : X_t = v\}$. The *return time* of v is $R_\alpha(v) = \min\{t \geq 1 : X_t = v \mid X_0 = v\}$. When v is understood, we simply write H_α and R_α. We write H and R for the hitting time and return time in a standard random walk.

3.2   Theorem 1

Before stating Theorem 1, we make the useful observation that we can split the α-random walk into two independent parts: (1) the portion preceding the first jump is the beginning of a standard random walk, and (2) the portion following the first jump is an α-random walk independent of the first portion. The probability that the first jump occurs at time t is $(1 - \alpha)^{t-1}\alpha$, i.e., the first jump time J is a geometric random variable with parameter α, independent of the nodes visited by the walk. Then we can model the α-random walk as follows: (1) start a standard random walk, (2) independently choose the first jump time J from a geometric distribution, and (3) at time J begin a new α-random walk. Hence we can express the return time and hitting time of v recursively:

$$R_\alpha = \begin{cases} R & \text{if } R < J \\ J + H'_\alpha & \text{otherwise} \end{cases}, \qquad H_\alpha = \begin{cases} H & \text{if } H < J \\ J + H'_\alpha & \text{otherwise} \end{cases}. \tag{1}$$

Here H'_α is an independent copy of H_α. It is convenient to abstract from our specific setting and state Theorem 1 about general random variables of this form.

---

[2] This is a technical condition that can be resolved in a variety of ways, for example, by adding self-loops to nodes with no outlinks.

Theorem 1. *Let R and H be independent, nonnegative, integer-valued random variables, and let J be a geometric random variable with parameter α. Define R_α and H_α as in (1). Then,*

(i) $E[R_\alpha] = \Pr[R \geq J]\left(\frac{1}{\alpha} + E[H_\alpha]\right)$,

(ii) $E[H_\alpha] = \frac{1}{\alpha} \cdot \frac{\Pr[H \geq J]}{\Pr[H < J]}$,

(iii) $E[R_\alpha] = \frac{1}{\alpha} \cdot \frac{\Pr[R \geq J]}{\Pr[H < J]}$.

Part (i) relates expected return time to expected hitting time: $\Pr[R \geq J]$ is the probability that the walk does not return before jumping. On the web, for example, we expect $\Pr[R \geq J]$ to be close to 1 for most pages, so the two measures are roughly equivalent. However, pages attempting to optimize PageRank can drive $\Pr[R \geq J]$ much lower, achieving an expected return time that is much lower than expected hitting time.

For parts (ii) and (iii), we adopt the convention that $\Pr[H < J] = 0$ implies $E[H_\alpha] = E[R_\alpha] = \infty$, corresponding to the case when v is not reachable from any node with positive restart probability. To gain some intuition for part (ii) (part (iii) is similar), we can think of the random walk as a sequence of independent explorations from the restart distribution "looking" for node v. Each exploration succeeds in finding v with probability $\Pr[H < J]$, so the expected number of explorations until success is $1/\Pr[H < J]$. The expected number of steps until an exploration is terminated by a jump is $1/\alpha$, so a rough estimate of hitting time is $\frac{1}{\alpha} \cdot \frac{1}{\Pr[H<J]}$. Of course, this is an overestimate because the final exploration is cut short when v is reached, and the expected length of an exploration conditioned on not reaching v is slightly shorter than $1/\alpha$. It turns out that $\Pr[H \geq J]$ is exactly the factor needed to correct the estimate, due to the useful fact about geometric random variables[3] stated in Lemma 1. We stress that the expected hitting time of v in the α-random walk is completely determined by $\Pr[H < J]$, the probability that a given exploration succeeds; this will serve as our numeric measure of reputation.

Lemma 1. *Let X and J be independent random variables such that X is nonnegative and integer-valued, and J is a geometric random variable with parameter α. Then $E[\min(X, J)] = \frac{1}{\alpha}\Pr[X \geq J]$.*

Lemma 1 is proved in the appendix.

Proof (Theorem 1). We rewrite $R_\alpha = \min(R, J) + I\{R \geq J\}H'_\alpha$, where $I\{R \geq J\}$ is the indicator variable for the event $R \geq J$. Note that $I\{R \geq J\}$ and H'_α are independent. Then, using linearity of expectation and Lemma 1,

---

[3] We mentioned that Theorem 1 can be derived from a result about regenerative stochastic processes [15]. In fact, Theorem 1 captures most of the generality; to write recurrences as in (1), the process need not be Markovian, it is only necessary that the process following a restart is a replica of the original. The only nongeneral assumption made is that J is a geometric random variable; this simplifies the conclusions.

$$E[R_\alpha] = E[\min(R, J)] + \Pr[R \geq J] E[H'_\alpha]$$
$$= \frac{1}{\alpha} \Pr[R \geq J] + \Pr[R \geq J] E[H_\alpha]$$
$$= \Pr[R \geq J] \left(\frac{1}{\alpha} + E[H_\alpha]\right).$$

This proves part (i). For part (ii), we take R to be a copy of H in part (i), giving

$$E[H_\alpha] = \Pr[H \geq J] \left(\frac{1}{\alpha} + E[H_\alpha]\right).$$

Solving this expression for $E[H_\alpha]$ gives (ii). Part (iii) is obtained by substituting (ii) into (i). □

4 Manipulation-Resistance

In this section we develop a reputation system based on hitting time, and quantify the extent to which an individual can tamper with reputations. It is intuitively clear that node u cannot improve its own hitting time by placing outlinks, but we would also like to limit the damage that u can cause to v's reputation. Specifically, u should only be able to damage v's reputation if u was responsible for v's reputation in the first place. Furthermore, u should not have a great influence on the reputation of too many others. To make these ideas precise, we define reputation using $\Pr[H < J]$ instead of $E[H_\alpha]$. By Theorem 1, either quantity determines the other — they are roughly inversely proportional — and $\Pr[H < J]$ is convenient for reasoning about manipulation.

Definition 1. *Let* $\mathrm{rep}(v) = \Pr[H(v) < J]$ *be the* reputation *of* v.

In words, $\mathrm{rep}(v)$ is the probability that a random walk hits v before jumping. Of all walks that reach v before jumping, an attacker u can only manipulate those that hit u first. This leads to our notion of influence.

Definition 2. *Let* $\mathrm{infl}(u, v) = \Pr[H(u) < H(v) < J]$ *be the* influence of u on v.

Definition 3. *Let* $\mathrm{infl}(u) = \sum_v \mathrm{infl}(u, v)$ *be the* total influence *of* u.

When the graph G is not clear from context, we write these quantities as $\Pr_G[\cdot]$, $\mathrm{rep}_G(\cdot)$ and $\mathrm{infl}_G(\cdot, \cdot)$ to be clear. To quantify what can change when u manipulates outlinks, let $\mathcal{N}_u(G)$ be the set of all graphs obtained from G by the addition or deletion of edges originating at u. It is convenient to formalize the intuition that u has no control over the random walk until it hits u for the first time.

Definition 4. *Fix a graph* G *and node* u. *We say that an event* A *is* u-invariant *if* $\Pr_G[A] = \Pr_{G'}[A]$ *for all* $G' \in \mathcal{N}_u(G)$. *If* A *is* u-invariant, *we also say that the quantity* $\Pr[A]$ *is* u-invariant.

Lemma 2. *An event A is u-invariant if the occurrence or non-occurrence of A is determined by time $H(u)$.*

Lemma 2 is proved in the appendix. With the definitions in place, we can quantify how much u can manipulate reputations.

Theorem 2. *For any graph $G = (V, E)$ and $u, v \in V$,*

(i) $\mathrm{infl}(u, u) = 0$,
(ii) $\mathrm{infl}(u, v) \geq 0$,
(iii) $\mathrm{infl}(u, v) \leq \mathrm{rep}(u)$,
(iv) $\mathrm{infl}(u) \leq \dfrac{1}{\alpha}\mathrm{rep}(u)$.

Let $G' \in \mathcal{N}_u(G)$. Then

(v) $\mathrm{rep}_{G'}(v) = \mathrm{rep}_G(v) + \mathrm{infl}_{G'}(u, v) - \mathrm{infl}_G(u, v)$.

Parts (i)-(iv) bound the influence of u in terms of its reputation. Part (v) states that when u modifies outlinks, the change in v's reputation is equal to the change in u's influence on v. Substituting parts (i-iii) into part (v) yields some simple but useful corollaries.

Corollary 1. *Let $G' \in \mathcal{N}_u(G)$. Then*

(i) $\mathrm{rep}_{G'}(u) = \mathrm{rep}_G(u)$,
(ii) $\mathrm{rep}_{G'}(v) \geq \mathrm{rep}_G(v) - \mathrm{infl}_G(u, v)$,
(iii) $\mathrm{rep}_{G'}(v) \leq \mathrm{rep}_G(v) - \mathrm{infl}_G(u, v) + \mathrm{rep}_G(u)$.

No matter what actions u takes, it cannot alter its own reputation (part (i)). Nor can u damage the portion of v's reputation *not* due to u's influence (part (ii)). On the other hand, u may boost its influence on v, but its final influence cannot exceed its reputation (part (iii)).

Proof (Theorem 2). For the most part, these are simple consequences of the definitions. Parts (i) and (ii) are trivial:

$$\mathrm{infl}(u, u) = \Pr\left[H(u) < H(u) < J\right] = 0,$$

$$\mathrm{infl}(u, v) = \Pr\left[H(u) < H(v) < J\right] \geq 0.$$

For part (iii), a walk that hits u then v before jumping contributes equally to u's reputation and u's influence on v:

$$\mathrm{infl}(u, v) = \Pr\left[H(u) < H(v) < J\right] \leq \Pr\left[H(u) < J\right] = \mathrm{rep}(u).$$

Part (iv) uses the observation that not too many nodes can be hit after u but before the first jump. Let $L = |\{v : H(u) < H(v) < J\}|$ be the number of all such nodes. Then,

$$E\left[L\right] = E\left[\sum_v I\{H(u) < H(v) < J\}\right] = \sum_v \Pr\left[H(u) < H(v) < J\right] = \mathrm{infl}(u).$$

But L cannot exceed $J - \min(H(u), J)$, so

$$
\begin{aligned}
\mathrm{infl}(u) = E\,[L] &\leq E\,[J] - E\,[\min(H(u), J)] \\
&= E\,[J]\,(1 - \Pr\,[H(u) \geq J]) \qquad \text{(by Lemma 1)} \\
&= E\,[J]\,\Pr\,[H(u) < J] \\
&= \frac{1}{\alpha}\mathrm{rep}(u).
\end{aligned}
$$

For part (v), we split walks that hit v before jumping into those that hit u first and those that don't:

$$
\begin{aligned}
\mathrm{rep}_G(v) &= \Pr_G\,[H(v) < J] \\
&= \Pr_G\,[H(u) < H(v),\ H(v) < J] + \Pr_G\,[H(u) \geq H(v),\ H(v) < J] \\
&= \mathrm{infl}_G(u, v) + \Pr_G\,[H(u) \geq H(v),\ H(v) < J]
\end{aligned}
$$

The event $[H(u) \geq H(v),\ H(v) < J]$ is determined by time $H(u)$, and hence it is u-invariant. By the above, $\Pr\,[H(u) \geq H(v),\ H(v) < J]$ is equal to $\mathrm{rep}_G(v) - \mathrm{infl}_G(u, v)$, and repeating the calculation for G' gives $\mathrm{rep}_{G'}(v) = \mathrm{infl}_{G'}(u, v) + \mathrm{rep}_G(v) - \mathrm{infl}_G(u, v)$. $\square$

4.1   Manipulating the Rankings

The previous results quantify how much node u can manipulate reputation *values*, but often we are more concerned with how much u can manipulate the ranking, specifically, how far u can advance by manipulating outlinks only. The following two corollaries follow easily, and are proved in the appendix. Suppose $\mathrm{rep}_G(u) < \mathrm{rep}_G(v)$ and u manipulates outlinks to produce $G' \in \mathcal{N}_u(G)$. We say that u *meets* v if $\mathrm{rep}_{G'}(u) = \mathrm{rep}_{G'}(v)$, and u *surpasses* v if $\mathrm{rep}_{G'}(u) > \mathrm{rep}_{G'}(v)$.

Corollary 2. *Node u cannot surpass a node that is at least twice as reputable.*

Corollary 3. *Node u can meet or surpass at most $\frac{1}{\alpha\gamma}$ nodes that are more reputable than u by a factor of at least $(1 + \gamma)$.*

4.2   Reputation and Influence of Sets

We have discussed reputation and influence in terms of individual nodes for ease of exposition, but all of the definitions and results generalize when we consider the reputation and influence of sets of nodes. Let $U, W \subseteq V$, and recall that $H(W) = \min_{w \in W} H(w)$ is the hitting time of the set W. Then we define $\mathrm{rep}(W) = \Pr\,[H(W) < J]$ to be the reputation of W, we define $\mathrm{infl}(U, W) = \Pr\,[H(U) < H(W) < J]$ to be the influence of U on W, and we define $\mathrm{infl}(U) = \sum_{v \in V} \mathrm{infl}(U, \{v\})$ to be the total influence of U. With these definitions, exact analogues of Theorem 2 and its corollaries hold for any $U, W \subseteq V$, with essentially the same proofs. Note that U and W need not be disjoint, in which case it is possible that $H(U) = H(W)$. We omit further details.

4.3 Sybils

In online environments, it is often easy for a user to create new identities, called *sybils*, and use them to increase her own reputation, even without obtaining any new inlinks from non-sybils. A wide class of reputation systems is vulnerable to sybil attacks [5], and, in the extreme, hitting time can be heavily swayed as well. For example, if u places enough sybils so the random walk almost surely starts at a sybil, then adding links from each sybil to u ensures the walk hits u by the second step unless it jumps. In this fashion, u can achieve reputation almost $1 - \alpha$ and drive the reputation of all non-sybils to zero. We'll see that this is actually the *only* way that sybils can aid u, by gathering restart probability and funneling it towards u. So an application can limit the effect of sybils by limiting the restart probability granted to new nodes. In fact, applications of hitting time analogous to Personalized PageRank [22] and TrustRank [14] are already immune, since they place all of the restart probability on a fixed set of known or trusted nodes. Applications like web search that give equal restart probability to each node are more vulnerable, but in cases like the web the sheer number of nodes requires an attacker to place many sybils to have a substantial effect. This stands in stark contrast with PageRank, where one sybil is enough to employ the 2-cycle self-endorsement strategy and increase PageRank by several times [6].

To model the sybil attack, suppose $G' = (V \cup S, E')$ is obtained from G by a sybil attack launched by u. That is, the sybil nodes S are added, and links originating at u or inside S can be set arbitrarily. All other links must not change, with the exception that those originally pointing to u can be directed anywhere within $S \cup \{u\}$. Let q' be the new restart distribution, assuming that q' diverts probability to S but does not redistribute probability within V. Specifically, if $\rho = \sum_{s \in S} q'(s)$ is the restart probability allotted to sybils, we require that $q'(v) = (1 - \rho)q(v)$ for all $v \in V$.

Theorem 3. *Let $U = \{u\} \cup S$ be the nodes controlled by the attacker u, and let v be any other node in V. Then*

  (i) $\mathrm{rep}_{G'}(u) \leq \mathrm{rep}_{G'}(U) = (1 - \rho)\mathrm{rep}_G(u) + \rho,$
  (ii) $\mathrm{rep}_{G'}(v) \geq (1 - \rho)(\mathrm{rep}_G(v) - \mathrm{infl}_G(u, v)),$
  (iii) $\mathrm{rep}_{G'}(v) \leq (1 - \rho)(\mathrm{rep}_G(v) - \mathrm{infl}_G(u, v) + \mathrm{rep}_G(u)) + \rho.$

Compared with Corollary 1, the only additional effect of sybils is to diminish all reputations by a factor of $(1 - \rho)$, and increase the reputation of certain target nodes by up to ρ.

Proof (Theorem 3). We split the attack into two steps, first observing how reputations change when the sybils are added but no links are changed, then applying Theorem 2 for the step when only links change. Let G^+ be the intermediate graph where we add the sybils but do not change links. Assume the sybils have self-loops so the transition probabilities are well-defined. We can compute $\mathrm{rep}_{G^+}(U)$ by conditioning on whether $X_0 \in V$ or $X_0 \in S$, recalling that $\Pr[X_0 \in S] = \rho$.

$$\mathrm{rep}_{G^+}(U) = (1-\rho) \cdot \mathrm{Pr}_{G^+}\left[H(U) < J \mid X_0 \in V\right] + \rho \cdot \mathrm{Pr}_{G^+}\left[H(U) < J \mid X_0 \in S\right]$$
$$= (1-\rho) \cdot \mathrm{Pr}_G\left[H(u) < J\right] + \rho$$
$$= (1-\rho)\mathrm{rep}_G(u) + \rho.$$

In the second step, $\mathrm{Pr}_{G^+}\left[H(U) < J \mid X_0 \in V\right] = \mathrm{Pr}_G\left[H(u) < J\right]$ because hitting U in G^+ is equivalent to hitting u in G; all edges outside U are unchanged, and all edges to U originally went to u. Also the conditional distribution of X_0 given $[X_0 \in V]$ is equal to q, by our assumption on q'. The term $\mathrm{Pr}_{G^+}\left[H(U) < J \mid X_0 \in S\right]$ is equal to one, since $X_0 \in S$ implies $H(U) = 0 < J$. A similar calculation gives

$$\mathrm{rep}_{G^+}(v) = (1-\rho)\mathrm{rep}_G(v) + \rho \cdot \mathrm{Pr}_{G^+}\left[H(v) < J \mid X_0 \in S\right] = (1-\rho)\mathrm{rep}_G(v).$$

The term $\mathrm{Pr}_{G^+}\left[H(v) < J \mid X_0 \in S\right]$ vanishes because S is disconnected, so a walk that starts in S cannot leave. Another similar calculation gives $\mathrm{infl}_{G^+}(U,v) = (1-\rho)\mathrm{infl}_G(u,v)$. Finally, we complete the sybil attack, obtaining G' from G^+ by making arbitrary changes to edges originating in U, and apply Corollary 1 (the version generalized to deal with sets) to G^+. Parts (i-iii) of this theorem are obtained by direct substitution into their counterparts from Corollary 1. □

Theorem 3 can also be generalized to deal with sets.

5 Computing Hitting Time

To realize a reputation system based on hitting time, we require an algorithm to efficiently compute the reputation of all nodes. Theorem 1 suggests several possibilities. Recall that $\pi(v)$ is the PageRank of v. Then $E\left[R_\alpha(v)\right] = 1/\pi(v)$ can be computed efficiently for all nodes using a standard PageRank algorithm, and the quantity $\mathrm{Pr}\left[R(v) \geq J\right]$ can be estimated efficiently by Monte Carlo sampling. Combining these two quantities using Theorem 1 yields $E\left[H_\alpha(v)\right]$.

It is tempting to estimate the reputation $\mathrm{Pr}\left[H(v) < J\right]$ directly using Monte Carlo sampling. However, there is an important distinction between the quantities $\mathrm{Pr}\left[R(v) \geq J\right]$ and $\mathrm{Pr}\left[H(v) < J\right]$. We can get one sample of either by running a random walk until it first jumps, which takes about $1/\alpha$ steps. However $\mathrm{Pr}\left[H(v) < J\right]$ may be infinitesimal, requiring a huge number of independent samples to obtain a good estimate. On the other hand, $\mathrm{Pr}\left[R(v) \geq J\right]$ is at least α since the walk has probability α of jumping in the very first step. If self-loops are disallowed, we obtain a better lower bound of $1 - (1-\alpha)^2$, the probability the walk jumps in the first two steps. For this reason we focus on $\mathrm{Pr}\left[R(v) \geq J\right]$.

5.1 A Monte Carlo Algorithm

In this section we describe an efficient Monte Carlo algorithm to simultaneously compute hitting time for all nodes. To obtain accuracy ϵ with probability at least $1 - \delta$, the time required will be $O(\frac{\log(1/\delta)}{\epsilon^2 \alpha^2}|V|)$ in addition to the time of one PageRank calculation. The algorithm is:

1. Compute π using a standard PageRank algorithm.[4] Then $E\left[R_\alpha(v)\right]=1/\pi(v)$.
2. For each node v, run k random walks starting from v until the walk either returns to v or jumps. Let $y_v = \frac{1}{k}\cdot(\#$ of walks that jump before returning to $v)$.
3. Use y_v as an estimate for $\Pr\left[R(v)\geq J\right]$ in part (i) or (iii) of Theorem 1 to compute $E\left[H_\alpha(v)\right]$ or $\Pr\left[H(v)<J\right]$.

How many samples are needed to achieve high accuracy? Let $\mu = \Pr\left[R(v)\geq J\right]$ be the quantity estimated by y_v. We call y_v an (ϵ,δ)-approximation for μ if $\Pr\left[|y_v - \mu|\geq\epsilon\mu\right]\leq\delta$. A standard application of the Chernoff bound (see [21] p. 254) shows that y_v is an (ϵ,δ)-approximation if $k\geq(3\ln(2/\delta))/\epsilon^2\mu$. Using the fact that $\mu\geq\alpha$, it is sufficient that $k\geq(3\ln(2/\delta))/\epsilon^2\alpha$. Since each walk terminates in $\frac{1}{\alpha}$ steps in expectation, the total expected number of steps is no more than $\frac{3\ln(2/\delta)}{\epsilon^2\alpha^2}|V|$.

For massive graphs like the web that do not easily fit into main memory, it is not feasible to collect the samples in step 2 of the algorithm sequentially, because each walk requires random access to the edges, which is prohibitively expensive for data structures stored on disk. We describe a method from [8] to collect all samples simultaneously making efficient use of disk I/O.

Conceptually, the idea is to run all walks simultaneously and incrementally by placing tokens on the nodes recording the location of each random walk. Then we can advance all tokens by a single step in one pass through the entire graph. Assuming the adjacency list is stored on disk sorted by node, we store the tokens in a separate list sorted in the same order. Each token records the node where it originated to determine if it returns before jumping. Then in one pass through both lists, we load the neighbors of each node into memory and process each of its tokens, terminating the walk and updating y_v if appropriate, else choosing a random outgoing edge to follow and updating the token. Updated tokens are written to the end of a new unsorted token list, and after all tokens are processed, the new list is sorted on disk to be used in the next pass.

The number of passes is bounded by the walk that takes the longest to jump, which is not completely satisfactory, so in practice we can stop after a fixed number of steps t, knowing that the contribution of walks longer than t is nominal for large enough t, since $\Pr\left[R\geq J, J>t\right]\leq\Pr\left[J>t\right]=(1-\alpha)^t$, which decays exponentially.

References

[1] Aldous, D., Fill, J.: Reversible Markov Chains and Random Walks on Graphs. Monograph in Preparation,
    http://www.stat.berkeley.edu/users/aldous/RWG/book.html
[2] Avrachenkov, K., Litvak, N., Nemirovsky, D., Osipova, N.: Monte carlo methods in PageRank computation: When one iteration is sufficient. Memorandum 1754, University of Twente, The Netherlands (2005)

---

[4] PageRank algorithms are typically iterative and incur some error. Our analysis bounds the additional error incurred by our algorithm.

[3] Bianchini, M., Gori, M., Scarselli, F.: Inside PageRank. ACM Trans. Inter. Tech. 5(1), 92–128 (2005)

[4] Brin, S., Page, L.: The anatomy of a large-scale hypertextual Web search engine. Computer Networks and ISDN Systems 30(1-7), 107–117 (1998)

[5] Cheng, A., Friedman, E.: Sybilproof reputation mechanisms. In: P2PECON 2005: Proceeding of the 2005 ACM SIGCOMM workshop on Economics of peer-to-peer systems, pp. 128–132. ACM Press, New York (2005)

[6] Cheng, A., Friedman, E.: Manipulability of PageRank under sybil strategies. In: Proceedings of the First Workshop of Networked Systems (2006)

[7] Douceur, J.: The sybil attack. In: Druschel, P., Kaashoek, M.F., Rowstron, A. (eds.) IPTPS 2002. LNCS, vol. 2429, Springer, Heidelberg (2002)

[8] Fogaras, D., Rácz, B.: Towards fully personalizing PageRank. In: Leonardi, S. (ed.) WAW 2004. LNCS, vol. 3243, Springer, Heidelberg (2004)

[9] Friedman, E., Resnick, P., Sami, R.: Manipulation-resistant reputation systems. In: Nisan, N., Roughgarden, T., Tardos, E., Vazirani, V. (eds.) Algorithmic Game Theory, Cambridge University Press, Cambridge (to appear)

[10] Gade, K., Prakash, A.: Using transient probability distributions of random walk to estimate spam resistant authority scores. Unpublished manuscript (2007)

[11] Gyöngyi, Z., Berkhin, P., Garcia-Molina, H., Pedersen, J.: Link spam detection based on mass estimation. In: Proceedings of the 32nd International Conference on Very Large Databases, ACM, New York (2006)

[12] Gyöngyi, Z., Garcia-Molina, H.: Link spam alliances. In: Proceedings of the 31st International Conference on Very Large Databases, pp. 517–528. ACM, New York (2005)

[13] Gyöngyi, Z., Garcia-Molina, H.: Web spam taxonomy. In: First International Workshop on Adversarial Information Retrieval on the Web (2005)

[14] Gyöngyi, Z., Garcia-Molina, H., Pedersen, J.: Combating web spam with TrustRank. In: Proceedings of the 30th International Conference on Very Large Databases, pp. 576–587. Morgan Kaufmann, San Francisco (2004)

[15] Heidelberger, P.: Fast simulation of rare events in queueing and reliability models. ACM Trans. Model. Comput. Simul. 5(1), 43–85 (1995)

[16] Jeh, G., Widom, J.: SimRank: A measure of structural-context similarity. In: Proceedings of the Eighth ACM SIGKDD International Conference on Knowledge Discovery and Data Mining (2002)

[17] Kamvar, S.D., Schlosser, M.T., Garcia-Molina, H.: The eigentrust algorithm for reputation management in P2P networks. In: WWW 2003: Proceedings of the 12th international conference on World Wide Web, pp. 640–651. ACM Press, New York (2003)

[18] Langville, A.N., Meyer, C.D.: Deeper inside PageRank. Internet Mathematics 1(3), 335–380 (2004)

[19] Liben-Nowell, D., Kleinberg, J.: The link prediction problem for social networks. In: CIKM 2003. Proceedings of the 12th International Conference on Information and Knowledge Management (2003)

[20] Mason, K.: Detecting Colluders in PageRank - Finding Slow Mixing States in a Markov Chain. PhD thesis, Stanford University (2005)

[21] Mitzenmacher, M., Upfal, E.: Probability and Computing: Randomized Algorithms and Probabilistic Analysis. Cambridge University Press, New York (2005)

[22] Page, L., Brin, S., Motwani, R., Winograd, T.: The PageRank citation ranking: Bringing order to the web. Technical report, Stanford Digital Library Technologies Project (1998)

[23] Zhang, H., Goel, A., Govindian, R., Mason, K., Van Roy, B.: Making eigenvector-based reputation systems robust to collusion. In: Leonardi, S. (ed.) WAW 2004. LNCS, vol. 3243, Springer, Heidelberg (2004)

A   Proofs

A.1   Lemma 1

Proof. Recall that J is the time of the first success in a sequence of independent trials that succeed with probability α, so $\Pr[J > t] = (1 - \alpha)^t$, and $\Pr[J \leq t] = 1 - (1 - \alpha)^t$.

$$
\begin{aligned}
E[\min(X, J)] &= \sum_{t=0}^{\infty} \Pr[\min(X, J) > t] \\
&= \sum_{t=0}^{\infty} \sum_{x=0}^{\infty} \Pr[X = x] \Pr[\min(X, J) > t \mid X = x] \\
&= \sum_{x=0}^{\infty} \Pr[X = x] \sum_{t=0}^{\infty} \Pr[\min(x, J) > t] \qquad \text{(using independence)} \\
&= \sum_{x=0}^{\infty} \Pr[X = x] \sum_{t=0}^{x-1} \Pr[J > t] \\
&= \sum_{x=0}^{\infty} \Pr[X = x] \sum_{t=0}^{x-1} (1 - \alpha)^t \\
&= \sum_{x=0}^{\infty} \Pr[X = x] \frac{1 - (1 - \alpha)^x}{1 - (1 - \alpha)} \\
&= \sum_{x=0}^{\infty} \Pr[X = x] \frac{\Pr[J \leq x]}{\alpha} \\
&= \frac{1}{\alpha} \Pr[X \geq J]
\end{aligned}
$$

$\square$

A.2   Lemma 2

Proof. Let $G' \in \mathcal{N}_u(G)$. It is enough to show that $\Pr_G[A \cap [H(u) = t]] = \Pr_{G'}[A \cap [H(u) = t]]$ for all $t \geq 0$. Let $W_{u,t}$ be the set of all walks that first hit u at step t. Specifically, $W_{u,t} = \{w_0 \ldots w_t : w_t = u, \ w_i \neq u \text{ for } i < t\}$. For $w = w_0 \ldots w_t$, let $\Pr[w]$ be shorthand for the probability of the walk w:

$$
\Pr[w] = \Pr[X_0 = w_0] \Pr[X_1 = w_1 \mid X_0 = w_0] \ldots \Pr[X_t = w_t \mid X_{t-1} = w_{t-1}].
$$

Then for $w \in W_{u,t}$, the transition probabilities in the expression above are independent of u's outlinks, so $\Pr_G[w] = \Pr_{G'}[w]$. Finally, since A is determined by time $H(u)$, there is a function $I_A : W_{u,t} \rightarrow \{0, 1\}$ that indicates the occurrence or non-occurrence of A for each $w \in W_{u,t}$. Putting it all together,

$$\Pr_G[A \cap [H(u) = t]] = \Pr_G[H(u) = t] \Pr_G[A \mid H(u) = t]$$

$$= \sum_{w \in W_{u,t}} \Pr_G[w] I_A(w)$$

$$= \sum_{w \in W_{u,t}} \Pr_{G'}[w] I_A(w)$$

$$= \Pr_{G'}[A \cap [H(u) = t]]$$

$\square$

A.3   Corollary 2

Proof. Suppose $\text{rep}_G(v) \geq 2 \cdot \text{rep}_G(u)$, then $\text{rep}_{G'}(v) \geq \text{rep}_G(v) - \text{infl}_G(u, v) \geq \text{rep}_G(v) - \text{rep}_G(u) \geq 2 \cdot \text{rep}_G(u) - \text{rep}_G(u) = \text{rep}_G(u) = \text{rep}_{G'}(u)$.

A.4   Corollary 3

Proof. Let $A = \{v : \text{rep}_G(v) \geq (1 + \gamma)\text{rep}_G(u), \text{rep}_{G'}(v) \leq \text{rep}_{G'}(u)\}$ be the set of all nodes with reputation at least $(1 + \gamma)$ times the reputation of u that are met or surpassed by u. Then

$$\sum_{v \in A} \text{rep}_G(v) \geq |A|(1 + \gamma)\text{rep}_G(u),$$

$$\sum_{v \in A} \text{rep}_{G'}(v) \leq |A|\text{rep}_{G'}(u) = |A|\text{rep}_G(u),$$

so $\sum_{v \in A}(\text{rep}_G(v) - \text{rep}_{G'}(v)) \geq \gamma|A|\text{rep}_G(u)$. But by Corollary 1, $\text{rep}_G(v) - \text{rep}_{G'}(v) \leq \text{infl}_G(u, v)$, so

$$\gamma|A|\text{rep}_G(u) \leq \sum_{v \in A}(\text{rep}_G(v) - \text{rep}_{G'}(v)) \leq \sum_{v \in A} \text{infl}_G(u, v) \leq \text{infl}_G(u) \leq \frac{1}{\alpha}\text{rep}_G(u),$$

hence $|A| \leq \frac{1}{\alpha\gamma}$. $\square$

Using Polynomial Chaos to Compute the Influence of Multiple Random Surfers in the PageRank Model

Paul G. Constantine and David F. Gleich

Stanford University
Institute for Computational and Mathematical Engineering
{paul.constantine,dgleich}@stanford.edu

Abstract. The PageRank equation computes the importance of pages in a web graph relative to a single random surfer with a constant teleportation coefficient. To be globally relevant, the teleportation coefficient should account for the influence of all users. Therefore, we correct the PageRank formulation by modeling the teleportation coefficient as a random variable distributed according to user behavior. With this correction, the PageRank values themselves become random. We present two methods to quantify the uncertainty in the random PageRank: a Monte Carlo sampling algorithm and an algorithm based the truncated polynomial chaos expansion of the random quantities. With each of these methods, we compute the expectation and standard deviation of the PageRanks. Our statistical analysis shows that the standard deviation of the PageRanks are uncorrelated with the PageRank vector.

1 Introduction

In its purest form, the PageRank model ignores the text underlying pages on the web and creates an irreducible, aperiod Markov chain model for a hypothetical random surfer on the link structure of the web [1]. Each entry of the stationary distribution measures the global importance of a page.

The PageRank model, however, is not unique. A PageRank value depends upon a parameter α which controls how the putative random surfer "teleports" around the web. Upon visiting a website, the random surfer chooses an outlink uniformly at random with probability α and chooses a page according to a prior distribution with probability $1 - \alpha$. This paper focuses on the modeling assumptions for the value of α and suggests a new model for PageRank that fixes a modeling error in the original PageRank formulation.

To continue our discussion, we must define the PageRank model and establish some notation. Let $\mathbf{W}$ be an adjacency matrix for a web graph, $w_{i,j} = 1$ when node i links to node j. We set $\mathbf{P}$ to be a fully row-stochastic random walk transition matrix on $\mathbf{W}$. The matrix $\mathbf{P}$ has dangling nodes corrected in an arbitrary way (for example, see [2,3]) such that $\mathbf{Pe} = \mathbf{e}$ where $\mathbf{e}$ is the vector of all ones. Let $1 - \alpha$ be the teleportation probability and $\mathbf{v}$ be the personalization

A. Bonato and F.R.K. Chung (Eds.): WAW 2007, LNCS 4863, pp. 82–95, 2007.

distribution. The PageRank model requires that $0 \leq \alpha < 1$, $v_i \geq 0$, and $\mathbf{v}^T \mathbf{e} = 1$. With these definitions, the PageRank vector $\mathbf{x}(\alpha)$ is the unique eigenvector with $||\mathbf{x}(\alpha)||_1 = 1$ satisfying

$$\left[\alpha \mathbf{P}^T + (1-\alpha)\mathbf{v}\mathbf{e}^T\right] \mathbf{x}(\alpha) = \mathbf{x}(\alpha) \tag{1}$$

or equivalently [4,5] the solution of the linear system

$$\left(\mathbf{I} - \alpha \mathbf{P}^T\right) \mathbf{x}(\alpha) = (1-\alpha)\mathbf{v}. \tag{2}$$

The key error in the PageRank model is that it only accounts for a single surfer because it only permits a single value of α. The choice of α is quite mysterious. Most researchers take $\alpha = 0.85$ [6]. Recently, Avrachenkov et al. suggested choosing $\alpha = 1/2$ [7]. Their suggestion follows from graph theoretic properties of the PageRank solution vector as a function of α. If we believe the PageRank random surfer model, then α should be estimated from Internet usage logs, so that $\alpha = \mathrm{E}[A]$ where A is a random variable representing the teleportation parameter for each user. We are not aware of any studies that attempt to determine α using this methodology.

However, assuming $\alpha = \mathrm{E}[A]$ does not yield the "correct" PageRank vector This fact follows because in general $\mathrm{E}[\mathbf{x}(A)] \neq \mathbf{x}(\mathrm{E}[A])$. Intuitively, this issue arises because the PageRank model consolidates everyone into a single user. Appendix A demonstrates a formal counterexample. A more realistic model would consider that each user should have a small contribution to the final PageRank values.

To reiterate, computing $\mathbf{x}(\mathrm{E}[A])$ does not yield a PageRank vector that expresses all of the users. Instead, we propose using $\mathrm{E}[\mathbf{x}(A)]$ as a new PageRank vector that accurately models the underlying user population.

While our model for PageRank using a random parameter better represents the reality of random surfers, we would not expect the rankings generated by the model to be qualitatively different from those generated by the approximation of using $\alpha = \mathrm{E}[A]$. We expect $\mathrm{E}[\mathbf{x}(A)] \approx \mathbf{x}(\mathrm{E}[A])$ for "reasonable" distributions of A. Our results confirm this expectation, which justifies use of the PageRank vector as a global ranking for all users.

However, by modeling each component of the PageRank vector as a random variable, we gain a distinct advantage when quantifying the importance of a page. Namely, we can compute the *standard deviation* of each PageRank value with respect to the distribution of A. The standard deviation is a key tool in uncertainty quantification and allows us to examine the pages *most sensitive* to changes in PageRank based on the underlying distribution of A. In the results section, we employ the standard deviation of the PageRank vector to generate rankings that are uncorrelated with the original PageRank vector. Uncorrelated vectors are important because they provide additional useful input to a machine learning framework for generating a web search ranking function.

2   Choice of Distribution

One of the mysteries in the original PageRank model was the choice of α. In our new model, we replace α with a random variable A. Immediately, we face a new question: what should the distribution of A be? This question is much easier to answer! We estimate a value $\hat{\alpha}_i$ from usage statistics for each user i and compute the resulting discrete distribution. Unfortunately, this represents a daunting computational and statistical analysis task.

We do not attempt to determine the underlying discrete distribution and proceed to the limiting case where A is a continuous random variable with support in the interval $[0, 1]$. There are two distributions that potentially model the user behavior of interest: the uniform distribution over $[l, r]$ with $0 \leq l < r \leq 1$ and the Beta distribution with parameters a and b, $a, b \geq -1$. The density functions are

$$f_{U[l,r]}(x) = \frac{1}{r-l} I_{[l,r]}(x) \text{ and } f_{\text{Beta}(a,b)}(x) = \frac{x^b (1-x)^a}{\text{B}(a+1, b+1)} I_{[0,1]}(x),$$

where $\text{B}(x, y) = \frac{\Gamma(x)\Gamma(y)}{\Gamma(x+y)}$ and $\Gamma(x)$ is the Euler Γ function. The Beta density becomes the uniform density with $l = 0$ and $r = 1$ when $a = b = 0$. (This form of the Beta density may differ from other presentations. In particular, the Beta distribution implemented in Matlab assumes the uniform distribution when $a = b = 1$.)

3   A Consequence of the Modeling Change

Recall that our proposed change in the PageRank model is to replace the deterministic parameter α with a random variable A and to use $\text{E}[\mathbf{x}(A)]$ instead of $\mathbf{x}(\text{E}[A])$ as the PageRank vector. One attractive feature of this change is that using $\text{E}[\mathbf{x}(A)]$ incorporates more influence from longer paths in the graph. Appendix A derives (5),

$$\text{E}[\mathbf{x}(A)] = \sum_{n=0}^{\infty} \text{E}[A^n - A^{n+1}] \mathbf{P}^{T^n} \mathbf{v}.$$

The coefficient $\text{E}[A^n - A^{n+1}]$ expresses the *weight* placed on paths of length n in the graph. Following Baeza-Yates et al., we call these coefficients the *path damping coefficients* [8]. Figure 1 shows these coefficients as functions of n along with the path damping coefficients in the deterministic case. Appendix A, then, demonstrates that the TotalRank model proposed in that paper is equivalent to using $\text{E}[\mathbf{x}(A)]$ in our model with A distributed uniformly over $[0, 1]$.

4   Computing the Solution

Given the randomness in the PageRank model introduced by the random parameter A, our objective is to quantify the uncertainty in the solution $\mathbf{x}(A)$

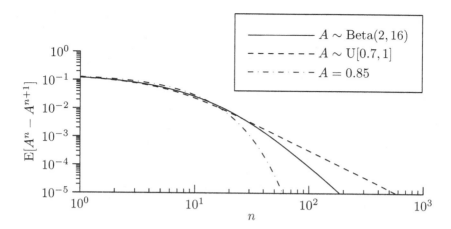

Fig. 1. Different choices of the underlying distribution of A have significant consequences for the influence of long paths in the final PageRank vector. From formula (5) the influence of a path of length n is $E[A^n - A^{n+1}]$ which we call the path damping coefficient. Assuming that A is uniform puts the most weight on long paths. The three distributions displayed in this plot satisfy $E[A] = 0.85$. All the methods put similar weight on paths up to length 10.

of the system. In concrete terms, this amounts to computing the mean of the stationary distribution $E[\mathbf{x}(A)]$ as well as its standard deviation $\text{Std}[\mathbf{x}(A)]$.

4.1 Monte Carlo Approach

One straightforward way to compute these quantities is to use a Monte Carlo approach. First generate M realizations of A from a chosen distribution, and then solve each resulting PageRank problem. With the M different realizations of $\mathbf{x}_i(A)$, $i = 1, \ldots, M$, we can compute unbiased estimates for $E[\mathbf{x}(A)]$ and $\text{Var}[\mathbf{x}(A)]$ with the formulas

$$E[\mathbf{x}(A)] \approx \frac{1}{M} \sum_{i=1}^{M} \mathbf{x}_i \equiv \hat{\mu}_\mathbf{x}, \quad \text{Std}[\mathbf{x}(A)] \approx \sqrt{\frac{1}{M-1} \sum_{i=1}^{M} (\mathbf{x}_i - \hat{\mu}_\mathbf{x})^2}$$

from [9]. Unfortunately, as with any Monte Carlo method, these estimates converge as $1/\sqrt{M}$ [10], which makes this approach prohibitively expensive for large systems such as the web graph.

4.2 The Polynomial Chaos Approach

A more efficient way to compute $E[\mathbf{x}(A)]$ and $\text{Std}[\mathbf{x}(A)]$ — and the one that we advocate in this paper — employs a technique known as the stochastic Galerkin method. This technique utilizes a specific representation of the random parameter A and random response vector $\mathbf{x}(A)$ called the polynomial chaos expansion

(PCE). The PCE expresses the random quantities as an infinite series of orthogonal polynomials that take a vector of random variables as arguments. This representation has its roots in the work of Wiener [11] who expressed a Gaussian process as an infinite series of Hermite polynomials. In the early 1990s, Ghanem and Spanos [12] truncated Wiener's representation to finitely many terms and used this truncated PCE as a primary component of their stochastic finite element method; this truncation made computations possible. In 2002, Xiu and Karniadakis [13] expanded this method to non-Gaussian processes via the more general Wiener-Askey scheme of orthogonal polynomials. The use of the stochastic Galerkin method has become fashionable in the uncertainty quantification community as a technique for measuring the effect of random inputs on partial differential equation models [14,15,16]. This paper presents a straightforward application of this technique to a linear system with a single random parameter.

To introduce the method, let $\{\Psi_k(\boldsymbol{\xi}(\omega))\}, k \in \mathbb{N}$ denote a set of orthogonal polynomials where $\boldsymbol{\xi}(\omega)$ is a vector of i.i.d. random variables. We use the dependence on ω to represent a random quantity. Assume $\{\Psi_k(\boldsymbol{\xi}(\omega))\}$ have the following properties:

$$\mathrm{E}[\Psi_0] = 1, \quad \mathrm{E}[\Psi_k] = 0 \text{ for } k > 0, \quad \mathrm{E}[\Psi_j\Psi_k] = \delta_{jk} \text{ for } j, k \geq 0$$

where δ_{jk} is the Kronecker delta. The PCE of a random quantity $u(\omega)$ is given by

$$u(\omega) = \sum_{k=0}^{\infty} u_k \Psi_k(\boldsymbol{\xi}(\omega))$$

where $\{u_i\}$ are the PCE coefficients. By the Cameron-Martin theorem [17], this series converges in an L_2 sense, i.e.

$$\mathrm{E}\left[\left(u - \sum_{k=0}^{N} u_k \Psi_k(\boldsymbol{\xi}(\omega))\right)^2\right] \rightarrow 0$$

as $N \rightarrow \infty$. The L_2 convergence of this expansion motivates truncating the series at a finite number of terms for the sake of computation. Thus we can approximate u with the finite series

$$u(\omega) \approx \sum_{k=0}^{N} u_k \Psi_k(\boldsymbol{\xi}(\omega)).$$

Then the problem of computing $u(\omega)$ transforms into the problem of finding the coefficients of its truncated PCE. From this point onward, we drop the explicit dependence on ω in our notation.

Since our particular model has only one random variable, we can fully account for this single random dimension by letting $\boldsymbol{\xi}$ have only one component, which we denote by $\boldsymbol{\xi} = \xi$. For the random parameter A and response quantity $\mathbf{x}$ in our model, we can write their respective PCEs as

$$A = \sum_{k=0}^{\infty} A_k \Psi_k(\xi), \qquad \mathbf{x} = \sum_{k=0}^{\infty} \mathbf{x}_k \Psi_k(\xi),$$

For $A \sim \text{Beta}(a, b)$, we can achieve exponential convergence in the coefficients $\mathbf{x}_k$ by choosing $\{\Psi_k\}$ to be the 1-D Jacobi polynomials with parameters a and b [13]. Note that when $a = b = 0$, $A \sim U[0, 1]$ and the Jacobi polynomials reduce to the Legendre polynomials.

Now we substitute these representations into our model,

$$\left(\mathbf{I} - \sum_{k=0}^{\infty} A_k \Psi_k(\xi) \mathbf{P}^T \right) \left(\sum_{k=0}^{\infty} \mathbf{x}_k \Psi_k(\xi) \right) = \left(1 - \sum_{k=0}^{\infty} A_k \Psi_k(\xi) \right) \mathbf{v}.$$

To make this problem amenable to computation, we truncate the PCEs to N terms,

$$\left(\mathbf{I} - \sum_{k=0}^{N} A_k \Psi_k(\xi) \mathbf{P}^T \right) \left(\sum_{k=0}^{N} \mathbf{x}_k \Psi_k(\xi) \right) = \left(1 - \sum_{k=0}^{N} A_k \Psi_k(\xi) \right) \mathbf{v}.$$

In the next section we perform a convergence study on the order of the expansion for our particular application, Fig. 2.

With the distribution of A known explicitly, we can solve directly for A_k. By multiplying both sides of the truncated PCE representation of A by Ψ_j and taking the expectation, the orthogonality of $\{\Psi_k\}$ gives the formula

$$A_j = \frac{\text{E}[A\Psi_j(\xi)]}{\text{E}[\Psi_j(\xi)^2]}, \quad j = 0, \ldots, N.$$

From this formula, we have that $A_0 = \text{E}[A]$ and $A_j = 0$ for $j \geq 2$. Thus our system reduces to

$$(\mathbf{I} - (A_0 + A_1 \Psi_1)\mathbf{P}^T) \left(\sum_{k=0}^{N} \mathbf{x}_k \Psi_k(\xi) \right) = (1 - (A_0 + A_1 \Psi_1))\mathbf{v}.$$

Multiplying both sides by Ψ_j and taking the expectation, the orthogonality of $\{\Psi_k\}$ leaves

$$\text{E}[\Psi_j^2]\mathbf{I} - \sum_{k=0}^{N} \text{E}[(A_0 + A_1 \Psi_1)\Psi_j \Psi_k]\mathbf{P}^T \mathbf{x}_k$$
$$= ((1 - A_0)\,\text{E}[\Psi_j] - A_1\,\text{E}[\Psi_j \Psi_1])\mathbf{v} \tag{3}$$

for $j = 0 \ldots N$. Therefore we have $N + 1$ coupled linear systems to solve for $\mathbf{x}_j$. Note that the dimension of this larger system is $N + 1$ times the dimension of $\mathbf{P}$.

Once we solve for the PCE coefficients $\mathbf{x}_j$, we can compute the mean of the PageRank vector,

$$E[\mathbf{x}] = E\left[\sum_{j=0}^{N} \mathbf{x}_j \Psi_j\right] = \mathbf{x}_0 \underbrace{E[\Psi_0]}_{=1} + \sum_{j=0}^{N} \mathbf{x}_j \underbrace{E[\Psi_i]}_{=0} = \mathbf{x}_0.$$

To compute the standard deviation, we first compute the variance.

$$\mathrm{Var}[\mathbf{x}] = E[(\mathbf{x} - E[\mathbf{x}])^2]$$

$$= E\left[\left(\left(\sum_{j=0}^{N} \mathbf{x}_j \Psi_j\right) - \mathbf{x}_0\right)^2\right]$$

$$= \sum_{j=1}^{N} \mathbf{x}_j^2 \, E[\Psi_j^2]; \quad \text{(by orthogonality)}$$

and then $\mathrm{Std}[\mathbf{x}] = \sqrt{\mathrm{Var}[\mathbf{x}]}$ is computed element-wise.

5   Datasets

Our experimental datasets came from four sources [18,19,20,21]. We downloaded and modified two datasets compressed using the Webgraph framework [22]. Table 1 summarizes our datasets.

From the webbase dataset, we extracted the web graph corresponding to the http://cs.stanford.edu host and computed the largest strongly connected component of this graph. We also computed and used the largest strongly connected component of the cnr-2000 graph. The wikipedia graph is discussed in Sect. 5.1, and the us2004 comes from [20]. Each of the graphs stanford-cs, cnr-2000, and wikipedia is a largest strongly connected component and has a natural random walk

$$\mathbf{P} = \mathbf{D}^{-1}\mathbf{W}, \tag{4}$$

Here $\mathbf{D}$ is the diagonal matrix of outdegrees for each node, and $\mathbf{Pe} = \mathbf{e}$. The us2004 graph was not strongly connected and we added self loops to all dangling nodes before computing $\mathbf{P}$ according to (4). The graphs for stanford-cs and wikipedia, prior to extracting the largest connected component, are available in the University of Florida Sparse Matrix Collection as Gleich/wb-cs-stanford and Gleich/wikipedia-20051105 [23].

5.1   Wikipedia

On a semi-regular basis, Wikipedia provides a dump of their database. We collected the dump from November 5, 2005 [21] and processed the results into a graph by identifying all links between Wikipedia articles in the text. We decided to remove many pages that were not articles because we wanted the results to be true to the underlying Encyclopedic nature of Wikipedia and felt that pages in the "User" and "User talk" categories did not meet that requirement. The categories we kept were "Category" and "Portal" because they represent overviews

Table 1. The datasets used in this paper vary in scale over three orders of magnitude. The term $id(v)$ is used to represent the indegree of a vertex. Each of the first three graphs listed is a strongly connected component from a larger graph. The final graph is not strongly connected and has dangling nodes adjusted by adding a self-loop.

Name	$|V|$	$|E|$	max $id(v)$	Source
stanford-cs	2,759	13,895	340	[18,22]
cnr-2000	112,023	1,646,332	18,235	[19,22]
wikipedia	1,103,453	18,245,140	71,524	Sect. 5.1
us2004	6,411,252	23,940,956	116,393	[20]

N	stanford-cs	cnr-2000
0	3.29×10^{-2}	3.20×10^{-2}
1	1.48×10^{-4}	1.47×10^{-4}
2	7.56×10^{-8}	7.54×10^{-8}
3	4.46×10^{-12}	4.47×10^{-12}
4	2.31×10^{-17}	2.75×10^{-17}

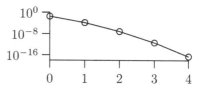

Values of $\left\| \mathrm{E}[\mathbf{x}_{(N+1)}(A)] - \mathrm{E}[\mathbf{x}_{(N)}(A)] \right\|_1$

Fig. 2. The order study for the polynomial chaos expansion shows that a fourth order expansion is sufficient and that the convergence of the expansion is independent of the graph size; the plots are indistinguishable and we have plotted the data from cnr-2000

of different areas of the Encyclopedia. Finally, we removed all pages in the graph not in the largest strongly connected component. The result of this processing is our wikipedia dataset.

6   Convergence Results

We conducted simple convergence studies with $A \sim \mathrm{Beta}(2, 16)$ on our two small datasets, stanford-cs and cnr-2000, that motivate choices used on the results for our larger datasets. From Figs. 2 and 3, we observed both the predicted

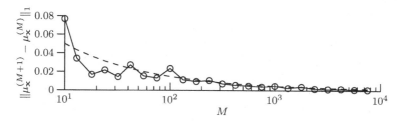

Fig. 3. Monte Carlo sampling theory predicts that the estimates converge proportional to $1/\sqrt{M}$. In this figure, we demonstrate this convergence on our problem. The dashed curve is $0.15/\sqrt{M}$ where the scaling was matched by hand to the underlying data.

exponential convergence of the PCE coefficients and the characteristic slow convergence of the Monte Carlo algorithm. Therefore, we use a fourth order PCE truncation ($N = 4$) in the our subsequent computations and do not consider Monte Carlo as a feasible computational algorithm. Appendix B describes the implementation for these experiments.

7   Results and Discussion

Tables 2-5 present our main results. In the experiments, we computed $\mathrm{E}[\mathbf{x}(A)]$ and $\mathrm{Std}[\mathbf{x}(A)]$ under the following modeling assumptions: (1) $A = 0.85$ is deterministic; (2) A has a Beta distribution with $a = 2$, $b = 16$; and (3) A is distributed uniformly over $[0.7, 1]$. We chose these two distributions because $\mathrm{E}[A] = 0.85$ in both cases. The two graphs used for experiments are wikipedia and us2004. In the remainder of this section, we will adopt the notation that $\hat{\mathbf{x}}_\alpha = \mathbf{x}(0.85)$, $\hat{\mathbf{x}}_{\mathrm{Beta}} = \mathrm{E}[\mathbf{x}(A)]$ where $A \sim \mathrm{Beta}(2, 16)$, and $\hat{\mathbf{x}}_U = \mathrm{E}[\mathbf{x}(A)]$ where $A \sim U[0.7, 1]$. Similarly, $\hat{\mathbf{s}}_{\mathrm{Beta}} = \mathrm{Std}[\mathbf{x}(A)]$ and $\hat{\mathbf{s}}_U = \mathrm{Std}[\mathbf{x}(A)]$ for the respective distributions.

Table 2 presents the time required for the major computational task in each evaluation. For the PageRank problem with a deterministic α, the major computational task is the iterative method used to solve the linear system (2). For the problem with a random variable A, the major computational task is solving the coupled linear systems for the coefficients of the PCE (3). The time required to compute the PCE coefficients is approximately 100 times larger than the time required to compute the PageRank vector. This implies that using our codes, we have an allowance of only 100 Monte Carlo samples before the PCE approach becomes more efficient. The computational codes for the PCE coefficients are not optimized. Therefore, using well known results and techniques from numerical linear algebra (for example [24]) will yield a substantial improvement in these computational times.

Next, we evaluated the top 10 pages for each of the ordering induced by $\hat{\mathbf{x}}_\alpha$, $\hat{\mathbf{x}}_{\mathrm{Beta}}$, and $\hat{\mathbf{x}}_U$. Unsurprisingly, these groups of pages are identical. We evaluate the difference between these three vectors in Tab. 3 using the 1-norm, ∞-norm, and Kendall-τ correlation coefficient. The results in the table clearly demonstrate that while $\mathrm{E}[\mathbf{x}(A)] \neq \mathbf{x}(\mathrm{E}[A])$, $\mathrm{E}[\mathbf{x}(A)] \approx \mathbf{x}(\mathrm{E}[A])$. From the strong τ correlation coefficients, the rankings induced by these vectors are nearly indistinguishable. These results justify using a deterministic approximation of PageRank to the underlying model where A is a random variable.

The next set of our results concerns the standard deviation of the PageRank vector with respect to the distribution of A. Table 4 displays the top 10 pages in each graph according to the PageRank vector and the top 10 pages with highest standard deviation under both distributions. The results first show that countries and years make up the most important pages in Wikipedia according to the PageRank model. Next, the pages with highest standard deviation are the category pages if A is sampled from a Beta distribution. This

Table 2. The time required to solve the major computational task in each model

	wikipedia	us2004
$A = \alpha = 0.85$	60.4 sec.	25.2 sec.
$A \sim \text{Beta}(2, 16)$	2210 sec.	3248 sec.
$A \sim U[0.7, 1.0]$	1936 sec.	2712 sec.

Table 3. In this table, $\hat{\mathbf{x}}_\alpha$ represents $\mathbf{x}(0.85)$, $\hat{\mathbf{x}}_U$ represents $E[\mathbf{x}(A)]$, $A \sim U[0.7, 1]$, and $\hat{\mathbf{x}}_{\text{Beta}} = E[\mathbf{x}(A)], A \sim \text{Beta}(2, 16)$. We present the difference between each of these vectors in the 1-norm, ∞-norm, and Kendall-τ correlation coefficient. These results show that the PageRank vector with a deterministic α is extremely close to the expected PageRank vectors for random variable A. In particular, the τ results indicate the rankings induced by each of the vectors are almost identical.

	wikipedia			us2004		
$\mathbf{y}, \mathbf{z}$	$\|\mathbf{y} - \mathbf{z}\|_1$	$\|\mathbf{y} - \mathbf{z}\|_\infty$	$\tau(\mathbf{y}, \mathbf{z})$	$\|\mathbf{y} - \mathbf{z}\|_1$	$\|\mathbf{y} - \mathbf{z}\|_\infty$	$\tau(\mathbf{y}, \mathbf{z})$
$\hat{\mathbf{x}}_\alpha, \hat{\mathbf{x}}_U$	0.04996	0.00018	0.99997	0.04996	0.00013	0.99999
$\hat{\mathbf{x}}_\alpha, \hat{\mathbf{x}}_{\text{Beta}}$	0.03211	0.00012	0.99702	0.03209	8×10^{-5}	0.99872
$\hat{\mathbf{x}}_U, \hat{\mathbf{x}}_{\text{Beta}}$	0.01792	5×10^{-5}	0.99705	0.01807	5×10^{-5}	0.99873

implies that these pages have the highest uncertainty in their ranking. A machine learning framework could theoretically use this additional information with user reviews of pages to generate more accurate rankings. Alternatively, pages with high standard deviation might make good suggestions if a search algorithm got a second chance to present results, given that the first set were unsatisfactory. In the second case, category pages are a logical set of suggestions. We were unable to determine any pattern to the pages with highest standard deviation if A is sampled from a uniform distribution. For the us2004 graph, the pages with highest PageRank tend to have high standard deviation as well.

We now attempt to analyze the standard deviation vectors using the Kendall-τ ranking correlation. Table 5 presents the correlation coefficients. The major result of this table is that the standard deviation vector for both distributions on the wikipedia graph is uncorrelated with any of the PageRank vectors whereas the standard deviation for both distributions on the us2004 graph are anti-correlated. When we evaluate the rank correlation in the us2004 graph with dangling nodes removed, then the rank correlation is positive. We believe that the difference between these results is a consequence of the structural differences between the graphs. The wikipedia graph is the largest strongly connected component of a larger graph and there are no dangling nodes. We believe that the dangling nodes in the us2004 graph are the cause of the strong anti-correlation between the standard deviation and the PageRank vector. The considerable change in the rank-correlation after removing the dangling nodes supports this hypothesis.

Table 4. This table compares the 10 best pages from each of the two experimental graphs with the 10 pages of highest standard deviation under each distributions for the random variable A. While each set of pages is significantly different in the wikipedia graph, the corresponding sets for the us2004 graph are similar.

wikipedia		
Top 10 by PageRank	**Largest std for Beta(2, 16)**	**Largest std for $U[0.7, 1.0]$**
United States	Category:Wikiportals	2000
Race (U.S. Census)	Category:Politics	Square kilometer
United Kingdom	Category:Categories	Population density
France	Category:Culture	Race (U.S. Census)
2005	Category:Geography	Per capita income
2004	Category:Countries	Poverty line
2000	Category:Human societies	Census
Canada	Race (U.S. Census)	Marriage
England	Category:Categories by country	Square mile
Category:Categories by country	Category:North American countries	Category:Categories by country

us2004		
Top 10 by PageRank	**Largest std for Beta(2, 16)**	**Largest std for $U[0.7, 1.0]$**
lxr.linux.no/	lxr.linux.no/	lxr.linux.no/
kernel.org/	kernel.org/	kernel.org/
examples.oreilly.com/linuxdrive2/	examples.oreilly.com/linuxdrive2/	examples.oreilly.com/linuxdrive2/
www.fsmlabs.com/[...]/openrtlinux/	www.fsmlabs.com/[...]/openrtlinux/	www.fsmlabs.com/[...]/openrtlinux/
www.datadosen.se/jalbum	www.datadosen.se/jalbum	www.datadosen.se/jalbum
linguistics.buffalo.edu/ssila/index.htm	www.iub.edu/	www.indiana.edu/copyright.html
www.iub.edu/	www.indiana.edu/&	www.indiana.edu/
www.uiowa.edu/ ournews/index.html	www.indiana.edu/copyright.html	registrar.indiana.edu/[...]/index.html
catalog.arizona.edu/	www.indiana.edu/	www.uky.edu/
validator.w3.org/check/referer	www.uiowa.edu/ournews/index.html	www.arizona.edu/

Table 5. This table shows the Kendall-τ ranking correlation coefficient between the standard deviation of the PageRank vector under both distributions for A and the expectations of the PageRank vector. See the text for the interpretation of the vectors in the table. For both the uniformly and Beta distributed random variables, the standard deviation of the wikipedia vector is uncorrelated with the PageRank vector itself. In contrast, the standard deviation of the us2004 vectors are anti-correlated with the PageRank vectors. The label us2004nd indicates the vector from the us2004 graph restricted to the non-dangling pages. On the non-dangling pages, we observe a more mild positive correlation between the PageRank vector and the standard deviation.

	wikipedia	us2004	us2004nd
$\tau(\hat{s}_U, \hat{x}_\alpha)$	-0.0032	-0.7846	0.4611
$\tau(\hat{s}_U, \hat{x}_U)$	-0.0032	-0.7846	0.4611
$\tau(\hat{s}_U, \hat{x}_{\text{Beta}})$	-0.0020	-0.7851	0.4625
$\tau(\hat{s}_{\text{Beta}}, \hat{x}_\alpha)$	0.0488	-0.7909	0.5920
$\tau(\hat{s}_{\text{Beta}}, \hat{x}_U)$	0.0488	-0.7909	0.5920
$\tau(\hat{s}_{\text{Beta}}, \hat{x}_{\text{Beta}})$	0.0021	-0.7918	0.5933

8   Conclusions and Future Work

The PageRank model for computing a global importance vector over web pages makes a critical modeling error by assuming that all users can be represented by a single teleportation parameter. We propose a new model for PageRank where the teleportation coefficient is a random variable supported on the interval $[0, 1]$.

Using a truncated polynomial chaos expansion to represent the random teleportation coefficient and PageRank, we compute the expectation and standard deviation of the PageRank vector for two distributions of the random variable in the PageRank model. Our results indicate two important conclusions. First, the expectation of the PageRank model assuming a random variable for the teleportation coefficient yields almost the same ranking as the PageRank model assuming a deterministic teleportation coefficient. This result justifies using the deterministic approximation as a global ranking vector for all users. Second, the standard deviation of the PageRank vector can be uncorrelated with the PageRank vector itself.

There is a significant amount of remaining work to fully investigate the use of the new model for PageRank. First, we need to investigate other distributions for the teleportation parameter, particularly distributions based on statistical analysis of actual usage data. Also, we suspect that the time required to compute the PCE coefficients can be significantly reduced. Additionally, we need to continue to investigate the impact of dangling nodes and strongly connected components on the standard deviation of the PageRank. We hypothesize that there is a significant relationship between these factors.

References

1. Page, L., Brin, S., Motwani, R., Winograd, T.: The PageRank citation ranking: Bringing order to the web. Technical Report 1999-66, Stanford University (1999)
2. Jeh, G., Widom, J.: Scaling personalized web search. In: Proceedings of the 12th international conference on the World Wide Web, Budapest, Hungary, pp. 271–279. ACM, New York (2003)
3. Kamvar, S.D., Haveliwala, T.H., Manning, C.D., Golub, G.H.: Extrapolation methods for accelerating PageRank computations. In: Proceedings of the 12th international conference on the World Wide Web, pp. 261–270. ACM Press, New York (2003)
4. Arasu, A., Novak, J., Tomkins, A., Tomlin, J.: PageRank computation and the structure of the web: Experiments and algorithms. In: Proceedings of the 11th international conference on the World Wide Web (2002)
5. Del Corso, G.M., Gullí, A., Romani, F.: Fast PageRank computation via a sparse linear system. Internet Mathematics 2(3), 251–273 (2005)
6. Langville, A.N., Meyer, C.D.: Google's PageRank and Beyond: The Science of Search Engine Rankings. Princeton University Press (2006)
7. Avrachenkov, K., Litvak, N., Pham, K.S.: A singular perturbation approach for choosing PageRank damping factor. arXiv e-prints (2006)
8. Baeza-Yates, R., Boldi, P., Castillo, C.: Generalizing PageRank: Damping functions for link-based ranking algorithms. In: Proceedings of ACM SIGIR, Seattle, Washington, USA, pp. 308–315. ACM Press, New York (2006)
9. Rice, J.A.: Mathematical Statistics and Data Analysis, 2nd edn. Duxbury Press, Boston (1995)
10. Ripley, B.D.: Stochastic Simulation, 1st edn. Wiley, Chichester (1987)
11. Wiener, N.: The homogeneous chaos. American Journal of Mathematics 60, 897–936 (1938)

12. Ghanem, R.G., Spanos, P.D.: Stochastic Finite Elements: A Spectral Approach, 1st edn. Springer, New York (1991)
13. Xiu, D., Karniadakis, G.E.: The Wiener–Askey polynomial chaos for stochastic differential equations. SIAM J. Sci. Comput. 24(2), 619–644 (2002)
14. Babuška, I., Tempone, R., Zouraris, G.: Galerkin finite element approximations of stochastic elliptic differential equations. SIAM Journal of Numeical Analysis 42(2), 800–825 (2004)
15. Wan, X., Karniadakis, G.E.: An adaptive multi-element generalized polynomial chaos method for stochastic differential equations. Journal of Computational Physics 209, 617–642 (2005)
16. Maître, O.P.L., Knio, O.M., Najm, H.N., Ghanem, R.G.: A stochastic projection method for fluid flow. Journal of Computational Physics 173, 481–511 (2001)
17. Cameron, R.H., Martin, W.T.: The orthogonal development of non-linear functionals in series of fourier-hermite functionals. The Annals of Mathematics 48, 385–392 (1947)
18. Hirai, J., Raghavan, S., Garcia-Molina, H., Paepcke, A.: WebBase: a repository of web pages. Computer Networks 33(1-6), 277–293 (2000)
19. Boldi, P., Codenotti, B., Santini, M., Vigna, S.: UbiCrawler: A scalable fully distributed web crawler. Software: Practice & Experience 34(8), 711–726 (2004)
20. Thelwall, M.: A free database of university web links: Data collection issues. International Journal of Scientometrics, Informetrics and Bibliometrics 6/7(1) (2003)
21. Various: Wikipedia XML database dump from November 5, 2005 (November 2005), Accessed from http://en.wikipedia.org/wiki/Wikipedia:Database_download
22. Boldi, P., Vigna, S.: Codes for the world wide web. Internet Mathematics 2, 407–429 (2005)
23. Davis, T.: University of Florida sparse matrix collection. NA Digest, vol. 92(42), October 16, 1994, NA Digest, vol. 96(28), July 23, 1996, and NA Digest, vol. 97(23), June 7, 1997 (2007), http://www.cise.ufl.edu/research/sparse/matrices/
24. Golub, G.H., van Loan, C.F.: Matrix Computations (Johns Hopkins Studies in Mathematical Sciences). The Johns Hopkins University Press, Baltimore (1996)

A   A Counterexample

We demonstrate $E[\mathbf{x}(A)] \neq \mathbf{x}(E[A])$ with a counterexample. Set

$$P = \begin{pmatrix} 0 & 1/2 & 1/2 \\ 0 & 0 & 1 \\ 0 & 0 & 1 \end{pmatrix},$$

and $v = \begin{bmatrix} 1/3 & 1/3 & 1/3 \end{bmatrix}^T$. If A is a uniform random variable on $[0,1]$, then $E[A] = 1/2$ and

$$\mathbf{x}(E[A]) = \begin{bmatrix} 1/6 & 5/24 & 5/8 \end{bmatrix}^T.$$

For the random variable model, the computations are more complicated.

$$\begin{aligned} E[\mathbf{x}(A)] &= E\left[\sum_{n=0}^{\infty} (A^n \mathbf{P}^{T^n})(1-A)\mathbf{v} \right] \\ &= \sum_{n=0}^{\infty} (E[A^n] - E[A^{n+1}])\mathbf{P}^{T^n}\mathbf{v}. \end{aligned} \tag{5}$$

In the first step, we used the Neumann series for the inverse of the matrix $\mathbf{I} - A\mathbf{P}^T$. Fubini's theorem then justifies interchanging the sum and expectation because the inner quantity is always bounded and positive. The raw moments of the uniform distribution are $\mathrm{E}[A^n] = \frac{1}{n+1}$ and consequently,

$$\mathrm{E}[\mathbf{x}(A)] = \sum_{n=0}^{\infty} \left(\tfrac{1}{n+1} - \tfrac{1}{n+2} \right) \mathbf{P}^{T^n} \mathbf{v} = \sum_{n=0}^{\infty} \tfrac{1}{(n+1)(n+2)} \mathbf{P}^{T^n} \mathbf{v}.$$

For $n \geq 2$, $\mathbf{P}^{T^n} \mathbf{v} = \begin{bmatrix} 0 & 0 & 1 \end{bmatrix}^T$

$$\mathrm{E}[\mathbf{x}(A)] = \tfrac{1}{2}\mathbf{v} + \tfrac{1}{6}\mathbf{P}^{\mathbf{T}}\mathbf{v} + \begin{bmatrix} 0 & 0 & \sum_{n=2}^{\infty} \frac{1}{(n+1)(n+2)} \end{bmatrix}^T$$
$$= \begin{bmatrix} 1/6 & 7/36 & 23/36 \end{bmatrix}^T.$$

In this case, the first component of the vector is identical, but the second two components show a small change from the modeling difference.

B   Engineering Details

Our Matlab implementations, which we provide for download from http://www.stanford.edu/~dgleich/pagerankpce, fall into three categories: PageRank codes, polynomial chaos codes, and large scale linear system codes.

PageRank codes. Our PageRank codes use a Gauss-Seidel algorithm [24] to solve the linear system formulation of the PageRank problem. We wrote a mex function to implement the Gauss-Seidel iteration efficiently in a Matlab code. The convergence metric for the PageRank computation was

$$||\alpha\mathbf{P}^T\mathbf{x} + (1 - \alpha)\mathbf{v} - \mathbf{x}||_1 \leq \delta$$

where $\delta = 10^{-10}$ for the PageRank vectors discussed in the results section and $\delta = 10^{-8}$ for the PageRank vectors used to form the Monte Carlo approximation.

Polynomial chaos codes. In the polynomial chaos approach, we must integrate products of the basis polynomials. We computed these quantities exactly with Matlab's symbolic toolbox. We used the output of these symbolic computations when forming the large linear system to solve for the PCE coefficients $\mathbf{x}_j$.

Large scale linear systems. We employed two linear system solvers to compute the solution for the large systems generated by the polynomial chaos approach. For the results in Sect. 6 we solved the final linear systems using the SOR algorithm with $\omega = 1.05$. For the results in Sect. 7, Matlab did not have sufficient memory to construct the large linear system in memory. We represented these matrices implicitly as linear operators and used the Jacobi algorithm to solve the linear systems until the relative residual was smaller than 10^{-10}.

A Spatial Web Graph Model with Local Influence Regions*

W. Aiello[1], A. Bonato[2], C. Cooper[3], J. Janssen[4], and P. Prałat[4]

[1] University of British Columbia
Vancouver, Canada
aiello@cs.ubc.ca
[2] Wilfrid Laurier University
Waterloo, Canada
abonato@rogers.com
[3] King's College
London, UK
colin.cooper@kcl.ac.uk
[4] Dalhousie University
Halifax, Canada
janssen@mathstat.dal.ca,pralat@mathstat.dal.ca

Abstract. The web graph may be considered as embedded in a topic space, with a metric that expresses the extent to which web pages are related to each other. Using this assumption, we present a new model for the web and other complex networks, based on a spatial embedding of the nodes, called the *Spatial Preferred Attachment (SPA)* model. In the SPA model, nodes have influence regions of varying size, and new nodes may only link to a node if they fall within its influence region. We prove that our model gives a power law in-degree distribution, with exponent in $[2, \infty)$ depending on the parameters, and with concentration for a wide range of in-degree values. We also show that the model allows for edges that span a large distance in the underlying space, modelling a feature often observed in real-world complex networks.

1 Introduction

Current stochastic models for complex networks (such as those described in [1,2]) aim to reproduce a number of graph properties observed in real-world networks such as the web graph. On the other hand, experimental and heuristic treatments of real-life networks operate under the tacit assumption that the network is a visible manifestation of an underlying hidden reality. For example, it is commonly assumed that communities in a social network can be recognized as densely linked subgraphs, or that web pages with many common neighbours contain related topics. Such assumptions imply that there is an a priori community structure or relatedness measure of the nodes, which is reflected by the link structure of the graph.

---

* The authors gratefully acknowledge support from NSERC and MITACS grants.

A. Bonato and F.R.K. Chung (Eds.): WAW 2007, LNCS 4863, pp. 96–107, 2007.

A common method to represent relatedness of objects is by an embedding in a metric space, so that related objects are placed close together, and communities are represented by clusters of points. Following a common text mining technique, web pages are often represented as vectors in a word-document space. Using Latent Sematic Indexing, these vectors can then be embedded in a Euclidean *topic space*, so that pages on similar topics are located close together. Experimental studies [7] have confirmed that similar pages are more likely to link to each other. On the other hand, experiments also confirm a large amount of *topic drift*: it is possible to move to a completely different topic in a relatively short number of hops. This points to a model where nodes are embedded in a metric space, and the edge probability between nodes is influenced by their proximity, but edges that span a larger distance in the space are not uncommon.

The *Spatial Preferred Attachment* (*SPA*) model proposed in this paper combines the above considerations with the often-used *preferential attachment principle*: pages with high in-degree are more likely to receive new links. In the SPA model, each node is placed in space and surrounded by an *influence region*. The area of the influence region is determined by the in-degree of the node. Moreover, in each time-step all regions decrease in area as a function of time. A new node v can only link to an existing node u if v falls within the influence region of u. If v falls within the region of influence u, then v will link to u with probability p. Thus, the model is based on the preferential attachment principle, but only implicitly: nodes with high in-degree have a large region of influence, and therefore are more likely to attract new links.

A random graph model with certain similarities to the SPA model is the *geometric random graph*; see [8]. In that model, all influence regions have the same size, and the link probability is $p = 1$. Flaxman, Frieze, and Vera in [5] supply an interesting geometric model where nodes are embedded on a sphere, and the link probability is influenced by the relative positions of the nodes. This model is a generalization of a geometric preferential attachment models presented by the same authors in [4], which influenced our model.

There are at least three features that distinguish the SPA model from previous work. First, a new node can choose its links purely based on *local* information. Namely, the influence region of a node can be seen as the region where a web page is *visible*: only web pages that are close enough (in topic) to fall within the influence region will be aware of the give page, and thus have a possibility to link to it. Moreover, a new node links independently to each node visible to it. Consequently, the new node needs no knowledge of the *invisible* part of the graph (such as in-degree of other nodes, or total number of nodes or links) to determine its neighbourhood. Second, since a new node links to each visible node independently, the out-degree is not a constant nor chosen according to a pre-determined distribution, but arises naturally from the model. Third, the varying size of the influence regions allows for the occasional *long links*, edges between nodes that are spaced far apart. This implies a certain "small world" property.

We formally define the SPA model as follows. Let S be the surface of the sphere of area 1 in $\mathbb{R}^3$. For each positive real number $\alpha \leq 1$, and $u \in S$, define the *cap around u with area α* as

$$B_\alpha(u) = \{x \in S : ||x - u|| \leq r_\alpha\},$$

where $|| \cdot ||$ is the usual Euclidean norm, and r_α is chosen such that B_α has area α.

The SPA model has parameters $A_1, A_2, A_3, p \geq 0$ such that $p \leq 1$, $A_1 \leq 1$ and $A_2 > 0$. It generates stochastic sequences of graphs $(G_t : t \geq 0)$, where $G_t = (V_t, E_t)$, and $V_t \subseteq S$. Let $d^-(v, t)$ $(d^+(v, t))$ be the in-degree (out-degree) of node v in G_t. We define the *influence region* of node v at time $t \geq 1$, written $R(v, t)$, to be the cap around v with area

$$|R(v, t)| = \frac{A_1 d^-(v, t) + A_2}{t + A_3},$$

or $R(v, t) = S$ if the righ-hand-side is greater than 1.

The process begins at $t = 0$, with G_0 being the empty graph, and we let G_1 be just K_1. Time-step t, $t \geq 2$, is defined to be the transition between G_{t-1} and G_t. At the beginning of each time-step t, a new node v_t is chosen uniformly at random (*uar*) from S, and added to V_{t-1} to create V_t. Next, independently, for each node $u \in V_{t-1}$ such that $v_t \in R(u, t - 1)$, a directed edge (v_t, u) is created with probability p. Thus, the probability that a link (v_t, u) is added in time-step t equals $p|R(u, t - 1)|$.

Because new nodes choose independently whether to link to each visible node, and the size of the influence region of a node depends only on the edges from *younger* nodes, the distribution of the random graph G_n produced by the SPA model with parameters A_1, A_2, A_3, p is equivalent to the graph G_{n+A_3} produced by the SPA model with the same values for A_1, A_2, p, but with $A_3 = 0$, where the first A_3 nodes have been removed. Since the results presented in this paper do not depend on the first nodes, we will assume throughout that $A_3 = 0$.

Note that the model could be defined on any compact set of measure 1. However, if the set has non-empty boundary, the definiton of the influence regions should be adjusted. If higher dimensions are desired, S could be chosen to be the boundary of a hypersphere in $\mathbb{R}^k$ for some k. The results in Sections 2 and 3 will still hold, while Section 4 can be easily extended to this case.

We prove in Section 2 that with high probability a graph G_n generated by the SPA model has an in-degree distribution that follows a power law in-degree distribution with exponent $1 + \frac{1}{pA_1}$, with concentration up to n^{i_f}, where $i_f = \left(\frac{n}{\log^4 n}\right)^{pA_1/(6pA_1+2)}$. If $pA_1 = 10/11$, then the power law in-degree exponent is 2.1, the same as observed in the web graph (see, for example [2]). We also give a precise expression for the probability distribution of each individual node v_i, provided that $pA_1 < 1$. In Section 3, we show that, if $pA_1 < 1$, the number of edges of G_n is linear, and strongly concentrated around the mean, while if $pA_1 = 1$ the expected number of edges is $n \log n$. In Section 4 we explore a geometric version of the small world property. We show that the expected sum of (geometric) lengths of new edges added at time t in the SPA model is $\Theta(t^{2-b})$, where $b = 1 + \frac{1}{pA_1}$ is the exponent of the power law. For the in-degree power law exponent $b = 2.1$ commonly observed in the web graph, this expected sum of lengths is greater than the corresponding expected sum in a corresponding geometric random graph with equal-sized influence regions.

2 In-Degree Distribution

In the rest of the paper, $(G_t : t \geq 0)$ refers to a sequence of random graphs generated by the SPA model with parameters $A_1, A_2, A_3 = 0$, and p. In this section, we explore the in-degree of the nodes in G_n. We say that an event holds *asymptotically almost surely* (*aas*) if it holds with probability tending to one as $n \to \infty$; an event holds *with extreme probability* (*wep*) if it holds with probability at least $1 - \exp(-\Theta(\log^2 n))$ as $n \to \infty$. Let $N_{i,t}$ denote the number of nodes of in-degree i in G_t. For an integer $n \geq 0$, define

$$i_f = i_f(n) = \left(\frac{n}{\log^4 n}\right)^{pA_1/(6pA_1+2)}. \tag{1}$$

Our main result in this section is the following.

Theorem 1. *Fix $p \in (0, 1]$. Then for any $i \geq 0$,*

$$\mathbb{E}(N_{i,n}) = c_i n(1 + o(1)), \tag{2}$$

where

$$c_0 = \frac{1}{1 + pA_2}, \tag{3}$$

and for $1 \leq i \leq n$,

$$c_i = \frac{p^i}{1 + pA_2 + ipA_1} \prod_{j=0}^{i-1} \frac{jA_1 + A_2}{1 + pA_2 + jpA_1}. \tag{4}$$

For $i = 0, \ldots, i_f$, wep

$$N_{i,n} = c_i n(1 + o(1)). \tag{5}$$

Since $c_i = ci^{-(1+\frac{1}{pA_1})}(1 + o(1))$ for some constant c, this shows that for large i, the expected proportion $N_{i,n}/n$ follows a power law with exponent $1 + \frac{1}{pA_1}$, with concentration for all values of i up to i_f. The proof of the Theorem 1 is contained in the rest of this section.

2.1 Expected Value

The equations relating the random variables $N_{i,t}$ are described as follows. As G_1 consist of one isolated node, $N_{0,1} = 1$, and $N_{i,1} = 0$ for $i > 0$. For all $t > 0$, we derive that

$$\mathbb{E}(N_{0,t+1} - N_{0,t} \mid G_t) = 1 - N_{0,t}p\frac{A_2}{t}, \tag{6}$$

$$\mathbb{E}(N_{i,t+1} - N_{i,t} \mid G_t) = N_{i-1,t}p\frac{A_1(i-1) + A_2}{t} - pN_{i,t}\frac{A_1 i + A_2}{t}. \tag{7}$$

Recurrence relations for the expected values of $N_{i,t}$ can be derived by taking the expectation of the above equations. To solve these relations, we use the following lemma on real sequences, which is Lemma 3.1 from [2].

Lemma 1. *If (α_t), (β_t) and (γ_t) are real sequences satisfying the relation*

$$\alpha_{t+1} = \left(1 - \frac{\beta_t}{t}\right)\alpha_t + \gamma_t,$$

and $\lim_{t \to \infty} \beta_t = \beta > 0$ and $\lim_{t \to \infty} \gamma_t = \gamma$, then $\lim_{t \to \infty} \frac{\alpha_t}{t}$ exists and equals $\frac{\gamma}{1+\beta}$.

Applying this lemma with $\alpha_t = \mathbb{E}(N_{0,t})$, $\beta_t = pA_2$, and $\gamma_t = 1$ gives that $\mathbb{E}(N_{0,t}) = c_0 t + o(t)$ with c_0 as in (3). For $i > 0$, the lemma can be inductively applied with $\alpha_t = \mathbb{E}(N_{i,t})$, $\beta_t = p(A_1 i + A_2)$, and $\gamma_t = \mathbb{E}(N_{i-1,t})\frac{A_1(i-1)+A_2}{t}$ to show that $\mathbb{E}(N_{i,t}) = c_i t + o(t)$, where

$$c_i = c_{i-1}\frac{A_1(i-1) + A_2}{1 + p(A_1 i + A_2)}.$$

It is easy to verify that the expression for c_i as defined in (3) and (4) satisfies this recurrence relation.

2.2   Concentration

We prove concentration for $N_{i,t}$ when $i \leq i_f$ by using a relaxation of Azuma-Hoeffding martingale techniques. The random variables $N_{i,t}$ do not a priori satisfy the c-Lipschitz condition: it is possible that a new node may fall into many overlapping regions of influence. Nevertheless, we will prove that deviation from the c-Lipschitz condition occurs with exponentially small probability. The following lemma gives a bound for $|N_{i,t+1} - N_{i,t}|$ which holds with extreme probability.

Lemma 2. Wep *for all $0 \leq t \leq n - 1$ the following inequalities hold.*

i $|N_{i,t+1} - N_{i,t}| \leq 2(A_1 i + A_2)\log^2 n$, *for $0 \leq i \leq t$.*

ii $|N_{i,t+1} - N_{i,t}| \leq 2(A_1 i + A_2)$, *for $\log^2 n < i \leq t$.*

Proof. Fix t, let $i, j \leq t$, and let $X_j(i, t)$ denote the indicator variable for the event that v_j has degree i at time t *and* v_{t+1} links to v_j. Thus,

$$N_{i,t+1} - N_{i,t} = \sum_{j=1}^{t} X_j(i - 1, t) - \sum_{j=1}^{t} X_j(i, t),$$

and so

$$|N_{i,t+1} - N_{i,t}| \leq \max\left(\sum_{j=1}^{t} X_j(i-1,t), \sum_{j=1}^{t} X_j(i,t)\right). \tag{8}$$

Let $Z_j(i,t)$ denote the indicator variable for the event that v_{t+1} is chosen in the cap of area $(A_1 i + A_2)/t$ around node v_j. Clearly, if $X_j(i,t) = 1$, then $Z_j(i,t) = 1$ as well, so $X_j(i,t) \leq Z_j(i,t)$. Thus, to bound $|N_{i,t+1} - N_{i,t}|$ it suffices to bound the values of $Z(i,t)$, where

$$Z(i,t) = \sum_{j=1}^{t} Z_j(i,t).$$

The variables $Z_j(i,t)$ for $j = 1, \ldots, t$ are pairwise independent. To see this, we can assume the position of v_{t+1} to be fixed. Then, the value of $Z_j(i,t)$ depends only on the position of v_j. Since the position of each node is chosen independently and uniformly, the value of $Z_j(i,t)$ is independent from the value of any other $Z_{j'}(i,t)$ where $j \neq j'$. Therefore, $Z(i,t)$ is the sum of independent Bernouilli variables with probability of success equal to

$$\mathbb{P}(Z_j(i,t) = 1) = \frac{A_1 i + A_2}{t}.$$

Using Chernoff's inequalities (see, for instance Theorem 2.1 [6]), we can show that $Z(i,t) < A_1 i + A_2 + (A_1 i + A_2) \log^2 n < 2(A_1 i + A_2) \log^2 n$. and $Z(i,t) < 2(A_1 i + A_2)$ if $i > \log^2 n$. Using these bounds, the proof now follows since by (8),

$$|N_{i,t+1} - N_{i,t}| \leq \max(Z(i-1,t), Z(i,t)). \qquad \square$$

To sketch the technique of the proof of Theorem 1, we consider $N_{0,t}$, the number of nodes of in-degree zero. We use the supermartingale method of Pittel et al. [9], as described in [10].

Lemma 3. *Let $G_0, G_1, \ldots, G_n$ be a random graph process and X_t a random variable determined by $G_0, G_1, \ldots, G_t$, $0 \leq t \leq n$. Suppose that for some real β and constants γ_i,*

$$\mathbb{E}(X_t - X_{t-1}|G_0, G_1, \ldots, G_{t-1}) < \beta$$

and

$$|X_t - X_{t-1} - \beta| \leq \gamma_i$$

for $1 \leq t \leq n$. Then for all $\alpha > 0$,

$$\mathbb{P}\big(\text{For some } t \text{ with } 0 \leq t \leq n : X_t - X_0 \geq t\beta + \alpha\big) \leq \exp\left(-\frac{\alpha^2}{2\sum \gamma_j^2}\right).$$

Note that we use the concept of a stopping time in the proof of Lemma 3 to obtain a stronger result. Stopping times aid by showing that the bound for the deviation of X_n applies with the same probability for all of the X_t, with $t \leq n$.

Theorem 2. Wep *for every* $t, 1 \leq t \leq n$

$$N_{0,t} = \frac{t}{1 + A_2 p} + O(n^{1/2} \log^3 n).$$

Proof. We first transform $N_{0,t}$ into something close to a martingale. Consider the following real-valued function

$$H(x, y) = x^{pA_2} y - \frac{x^{1+pA_2}}{1 + pA_2} \tag{9}$$

(note that we expect $H(t, N_{0,t})$ to be close to zero). Let $\mathbf{w}_t = (t, N_{0,t})$, and consider the sequence of random variables $(H(\mathbf{w}_t) : 1 \leq i \leq n)$. The second-order partial derivatives of H evaluated at $\mathbf{w}_t$ are all $O(t^{pA_2-1})$. Therefore, we have

$$H(\mathbf{w}_{t+1}) - H(\mathbf{w}_t) = (\mathbf{w}_{t+1} - \mathbf{w}_t) \cdot \operatorname{grad} H(\mathbf{w}_t) + O(t^{pA_2-1}), \tag{10}$$

where "$\cdot$" denotes the scalar product and $\operatorname{grad} H(\mathbf{w}_t) = (H_x(\mathbf{w}_t), H_y(\mathbf{w}_t))$.

Observe that, from our choice of H,

$$\mathbb{E}(\mathbf{w}_{t+1} - \mathbf{w}_t \mid G_t) \cdot \operatorname{grad} H(\mathbf{w}_t) = 0,$$

since H was chosen so that $H(\mathbf{w})$ is constant along every trajectory $\mathbf{w}$ of the differential equation that approximates the recurrence relation (6).

Hence, taking the expectation of (10) conditional on G_t, we obtain that

$$\mathbb{E}(H(\mathbf{w}_{t+1}) - H(\mathbf{w}_t) \mid G_t) = O(t^{pA_2-1}).$$

From (10), noting that

$$\operatorname{grad} H(\mathbf{w}_t) = \left(pA_2 t^{pA_2-1} N_{0,t} - t^{pA_2}, t^{pA_2} \right),$$

and using Lemma 2 to bound the change in $N_{0,t}$, we have that *wep*

$$|H(\mathbf{w}_{t+1}) - H(\mathbf{w}_t)| \leq t^{pA_2} 2(A_1 i + A_2) \log^2 n + O(t^{pA_2}) = O(t^{pA_2} \log^2 n).$$

Now we may apply Lemma 3 to the sequence $(H(\mathbf{w}_t) : 1 \leq i \leq n)$, and symmetrically to $(-H(\mathbf{w}_t) : 1 \leq i \leq n)$, with $\alpha = n^{1/2+pA_2} \log^3 n$, $\beta = O(t^{pA_2-1})$ and $\gamma_t = O(t^{pA_2} \log^2 n)$, to obtain that wep

$$|H(\mathbf{w}_t) - H(\mathbf{w}_0)| = O(n^{1/2+pA_2} \log^3 n)$$

for $1 \leq t \leq n$. As $H(\mathbf{w}_0) = 0$, this implies from the definition (9) of the function H, that *wep*

$$N_{0,t} = \frac{t}{1 + pA_2} + O(n^{1/2} \log^3 n) \tag{11}$$

for $1 \leq t \leq n$ which finishes the proof of the theorem. □

We may repeat the argument as in the proof of Theorem 2 for $N_{i,t}$ with $i \geq 1$. We omit the details here, which will follow in the long version of the paper.

2.3   In-Degree of Given Node

In contrast to the large-scale behaviour of the degree distribution described in the previous subsection, here we focus on the distribution of the in-degree of an individual node. The indicator variable Y_t for the increase in $d^-(v,t)$ by receiving a link from v_{t+1} is Bernoulli $\mathrm{Be}(p(A_1 d^-(v,t) + A_2)/t)$. Thus,

$$\mathbb{E}(d^-(v,t+1)|G_t) = d^-(v,t) + \frac{p(A_1 d^-(v,t) + A_2)}{t}. \qquad (12)$$

This is very similar to the growth of the degree in the Preferential Attachment model as analized in [3]. As in the PA model, a "rich get richer" principle applies for the in-degrees, and the richer nodes are those that were born first. Theorem 2.1 of [3] can be used to obtain results on the concentration of $N_{i,t}$, but the methods employed in the previous sections give a stronger result.

The results on the distribution of $d^-(v,n)$ are summarized in parts (a) and (b) of the theorem below (use Theorem 2.2 of [3] with minor reworking). Part (c) will be discussed in the next section, and used to establish the concentration of the edges of G_t.

Theorem 3. *Let* $\omega = \log n$ *and let* $l^* = n^{\min\{pA_1, 1/2\}}/\omega^4$. *For* $0 < pA_1 < 1$,

(a) For $\omega^8 \leq j \leq (n - n/\omega)$ *and* $0 \leq l \leq l^*$ *or for* $(n - n/\omega) < j < n$ *and* $l = 0, 1$,

$$\mathbb{P}(d^-(v_j, n) = l) = (1 + O(1/\omega^2)) \left(\frac{n}{j}\right)^{pA_1} \left(1 - \left(\frac{n}{j}\right)^{pA_1} (1 + O(1/\omega^2))\right)^l.$$

(b) For $(n - n/\omega) < j < n$ *and* $l \geq 2$,

$$\mathbb{P}(d^-(v_j, n) = l) = O(l^{pA_1 - 1}/\omega^l).$$

(c) For all $K > 0$,

$$\mathbb{P}(\text{There exists } j \leq n : d^-(v_j, n) \geq K\omega^2 (n/j)^{pA_1}) = O\left(n^{-Ke^{-18}}\right).$$

Theorem 3(c) implies that *aas* the maximum in-degree of node v_j is at most $(n/j)^{pA_1} K\omega^2$. Conditional on this, (a) and (b) characterize the distribution of $d^-(v_j, n)$ for all $j \geq \omega^8$ when $pA_1 \leq 1/2$ and for $j \geq \omega^8 n^{pA_1 - 1/2}$ when $pA_1 > 1/2$.

3   The Number of Edges of G_t

We derive a concentration result for the number of edges in graphs generated by the SPA model. Let $M_t = |E_t|$, the number of edges in G_t, and let $m_t = \mathbb{E}(M_t)$. Then we have that

$$\mathbb{E}(M_{t+1} \mid M_t) = M_t + \sum_{j=1}^{t} p \frac{A_1 d^-(v_j, t) + A_2}{t} = M_t + \frac{pA_1 M_t}{t} + pA_2,$$

and so $m_1 = 0$, and for $t \geq 1$,

$$m_{t+1} = m_t \left(1 + \frac{pA_1}{t}\right) + pA_2.$$

The (first-order) solutions of this recurrence are

$$m_n \sim \begin{cases} \frac{pA_2}{1-pA_1} n, \ pA_1 < 1 \\ \\ n \log n, \ \ pA_1 = 1. \end{cases}$$

Theorem 4. *If $pA_1 < 1$, then aas the number of edges is concentrated around its expected value:*

$$M_n = m_n(1 + o(1)).$$

The following lemma (whose proof is left to the long version of the paper) is used in the proof of Theorem 4, and proves Theorem 3 (c).

Lemma 4. *For all v_j, $j > 0$ and $K > 0$,*

$$\mathbb{P}(d^-(v_j, n) \geq K \log^2 n(n/j)^{pA_1}) = O(n^{-Ke^{-18}}).$$

Proof of Theorem 4. We count the number of edges by counting the in-degree of nodes. Our approach is as follows: by Theorem 1 *wep* for $i \leq i_f$ the number of nodes $N_{i,n}$ of in-degree i at time n is concentrated. Let a be the solution of $(n/a)^{pA_1} = i_f$ and let $\omega' = (K \log^2 n)^{1/(pA_1)}$ be the solution of

$$\left(\frac{t}{a\omega'}\right)^{pA_1} K \log^2(n) = \left(\frac{n}{a}\right)^{pA_1},$$

where $K \geq 4e^{18}$. From Lemma 4, with probability $1 - O(n^{-3})$ no node $v \geq a\omega'$ has degree exceeding i_f. Let $\mu(n) = \sum_{i \leq i_f} \mathbb{E}N_{i,n}$, and let $\lambda(n) = \sum_{j=1}^{a\omega'} d^-(v_j, n)$. We prove, conditional on Lemma 4, that $\lambda(n) = o(m_n)$ and thus the number of edges is concentrated around m_n. We have that for $pA_1 < 1$.

$$\lambda(n) = \sum_{j=1}^{a\omega'} d^-(v_j, n)$$

$$\leq K\omega^2 \sum_{j=1}^{a\omega'} \left(\frac{n}{j}\right)^{pA_1}$$

$$= O(1/(1 - pA_1)) \log^{2/(pA_1)}(n) n^{pA_1} a^{1-pA_1}$$

$$= O(1/(1 - pA_1)) \log^{2/(pA_1)}(n) + 4(1 - pA_1)/(6pA_1 + 2) n^{\frac{7pA_1+1}{6pA_1+2}}$$

$$= o(n).$$

However, $\mu(t) \geq ct$ for some constant $c > 0$. $\qquad \square$

4   A Geometric Small World Property

In Section 2 it was shown that the number of nodes in a graph generated by the SPA model of in-degree zero in G_n is linear in n. Also, with positive probability a new node will land in an area of S not covered by any influence regions, and thus have out-degree zero. Therefore, the underlying undirected graph of G_n is not connected. In fact, we expect that for the majority of distinct pairs u, v, there will not be a directed path from u to v. Since this is a property also observed in the web graph, it does not detract from the SPA model, but rather indicates that we should consider another variable rather than diameter to indicate a "small world" property. Thus, we focus on the (geometric) distance, in S, spanned by the links.

For a pair of points $u, v \in S$. let $L(u, v)$ be the length of the shortest curve embedded in the surface of S that connects u and v. Define

$$L_t = \sum_{(v_t, v_i) \in E_t} L(v_t, v_i);$$

that is, L_t is the sum of the lengths of new edges added at time t in the SPA model. Note that L_t is a continuous random variable.

Theorem 5. *Suppose that $pA_1 > 2/3$. For the expectation of L_t,*

$$\mathbb{E}(L_t) = \Theta\left(t^{-\left(\frac{1-pA_1}{pA_1}\right)}\right).$$

To prove Theorem 5 we need the following lemma whose (straightforward) proof is omitted.

Lemma 5. *Let u be chosen uar from a cap with centre v and area α. If X is the distance between u and v, measured over the surface of S, then $\mathbb{E}(X) = \frac{2}{3}\sqrt{\frac{\alpha}{\pi}}$.*

Proof of Theorem 5. Define

$$Z_{j,t} = \begin{cases} L(v_t, v_j) & \text{if } (v_t, v_j) \in E_t \\ 0 & \text{else.} \end{cases}$$

Then $L_t = \sum_{j=1}^{t-1} Z_{j,t}$. Let $B_{t,j}$ be the event that $(v_t, v_j) \in E_t$. Then using Lemma 5 we have that

$$
\begin{aligned}
\mathbb{E}(Z_{j,t+1} \mid G_t) &= \mathbb{P}(B_{t,j})\mathbb{E}(Z_{j,t+1} \mid G_t, B_{t,j}) + \mathbb{P}(\overline{B_{t,j}})\mathbb{E}(Z_{j,t+1} \mid G_t, \overline{B_{t,j}}) \\
&= \mathbb{P}(B_{t,j})\mathbb{E}(L((v_{t+1}, v_j) \mid G_t) \\
&= \left(p\frac{A_1 d^-(v_j, t) + A_2}{t}\right)\left(\frac{2}{3}\sqrt{\frac{A_1 d^-(v_j, t) + A_2}{\pi t}}\right) \\
&= \frac{2p}{3\sqrt{\pi}}\left(\frac{A_1 d^-(v_j, t) + A_2}{t}\right)^{3/2},
\end{aligned}
$$

where the second last equality follows by Lemma 5 and the definition of the model, and the second equality follows from the definition of $Z_{j,t+1}$. Thus

$$\mathbb{E}(L_{t+1} \mid G_t) = \sum_{k=0}^{t} \sum_{\{j:d^-(v_j,t)=k\}} \mathbb{E}(Z_{j,t+1}|G_t) = \frac{2p}{3\sqrt{\pi}} \sum_{k=0}^{t} \left(\frac{A_1 k + A_2}{t}\right)^{3/2} N_{k,t}.$$

(13)

Taking expectations on both sides, and using that $c_k = ck^{-(1+\frac{1}{pA_1})}(1 + o(1))$, we have that

$$\mathbb{E}(L_{t+1}) = \frac{2p}{3\sqrt{\pi}} \sum_{k=0}^{t} \left(\frac{A_1 k + A_2}{t}\right)^{3/2} \mathbb{E}(N_{k,t})$$

$$= \frac{2p}{3\sqrt{\pi t}} \sum_{k=0}^{t} (A_1 k + A_2)^{3/2} c_k (1 + o(1))$$

$$= \frac{2pc}{3\sqrt{\pi t}} \int_0^t x^{1/2 - 1/(pA_1)} (1 + o(1)) dx$$

$$= \Theta(t^{1 - 1/(pA_1)}),$$

where the second equality follows by Theorem 1 (2). The last step is justified since it can be shown that the $o(1)$ term in the integrand is in fact $O(x^{-\epsilon})$ for some $\epsilon > 0$. □

Theorem 5 contrasts with the analogous result for graphs generated with a similar process to the SPA model, but where all influence regions have area d/t for $d > 0$ a constant. We call this a *threshold model*. In the threshold model, $\mathbb{E}(L_t)$ decreases much faster than for the SPA model with p large, such as when $p > 2/3$ and $A_1 = 1$. For example, if $pA_1 = 1$, then $\mathbb{E}(L_t) = O(1)$.

Theorem 6. *In the threshold model with areas of influence d/t, where d is a constant,*

$$\mathbb{E}(L_t) \sim ct^{-1/2}.$$

Proof. With the same notation as in the proof of Theorem 5 and using Lemma 5, we have that

$$\mathbb{E}(Z_{j,t+1} \mid G_t) = \mathbb{P}(B_{j,t+1})\mathbb{E}(L(v_{t+1}, v_j) \mid B_{j,t+1})$$

$$= \frac{2d}{3t} \sqrt{\frac{d}{\pi t}}.$$

Hence,

$$\mathbb{E}(L_{t+1} \mid G_t) = \sum_{i=1}^{t} \mathbb{E}(Z_{j,t+1}|G_t) = \frac{2d}{3} \sqrt{\frac{d}{\pi t}} = \Theta(t^{-1/2}).$$

Taking expectations completes the proof. □

5 Conclusions and Further Work

We have proved that graphs produced by the SPA model have some of the graph properties observed in real-world complex networks: a power law in-degree distribution, and constant average degree. In future work, we will investigate additional graph properties, such as the expected length of a directed path between two nodes (when such a path exists), expansion properties, and spectral values. We are also interested in aspects suggesting *self-similarity*: is it true that the subgraph induced by all nodes that fall in a certain compact region of the sphere S share some of the graph properties of the whole graph?

Several generalizations of this model may be proposed. An undirected version could be developed, where the link probability depends on the influence regions of both endpoints. In a more realistic model, both the addition of edges without adding a node and the deletion of edges and nodes should be incorporated. The effect of replacing S with other underlying geometric spaces, either with boundaries or of higher dimension, would be interesting to investigate.

Last but not least, a realistic spatial model gives the possibility for *reverse engineering* of real-life networks: given a real-life network and assuming a spatial graph model by which the network was generated, it should be possible to give reliable estimates about the positions of the nodes in space. This direction has important applications to web graph clustering and development of link-based similarity measures.

References

1. Bonato, A.: A survey of web graph models. Proceedings of Combinatorial and Algorithm Aspects of Networking (2004)
2. Chung, F.R.K., Lu, L.: Complex Graphs and Networks. American Mathematical Society, Providence (2006)
3. Cooper, C.: The age specific degree distribution of web-graphs. Combinatorics Probability and Computing 15, 637–661 (2006)
4. Flaxman, A., Frieze, A.M., Vera, J.: A geometric preferential attachment model of networks. Internet Mathematics 3, 187–205 (2006)
5. Flaxman, A., Frieze, A.M., Vera, J.: A geometric preferential attachment model of networks II (preprint)
6. Janson, S., Łuczak, T., Ruciński, A.: Random Graphs. Wiley, New York (2000)
7. Menczer, F.: Lexical and semantic clustering by Web links. JASIST 55(14), 1261–1269 (2004)
8. Penrose, M.: Random Geometric Graphs. Oxford University Press, Oxford (2003)
9. Pittel, B., Spencer, J., Wormald, N.: Sudden emergence of a giant k-core in a random graph. Journal of Combinatorial Theory, Series B 67, 111–151 (1996)
10. Wormald, N.: The differential equation method for random graph processes and greedy algorithms. In: Karoński, M., Prömel, H.J. (eds.) Lectures on Approximation and Randomized Algorithms, PWN, Warsaw, pp. 73–155 (1999)

Determining Factors Behind the PageRank Log-Log Plot

Yana Volkovich[1], Nelly Litvak[1], and Debora Donato[2]

[1] University of Twente, P.O. Box 217, 7500 AE, Enschede, The Netherlands
[2] Yahoo! Research, Barcelona Ocata 1, 1st floor, 08003, Barcelona Catalunya, Spain

Abstract. We study the relation between PageRank and other parameters of information networks such as in-degree, out-degree, and the
fraction of dangling nodes. We model this relation through a stochastic equation inspired by the original definition of PageRank. Further, we
use the theory of regular variation to prove that PageRank and in-degree
follow power laws with the same exponent. The difference between these
two power laws is in a multiplicative constant, which depends mainly on
the fraction of dangling nodes, average in-degree, the power law exponent, and the damping factor. The out-degree distribution has a minor
effect, which we explicitly quantify. Finally, we propose a ranking scheme
which does not depend on out-degrees.

Keywords: PageRank, Power laws, Ranking algorithms, Stochastic
equations, Web graph, Wikipedia.

1 Introduction

Originally created for Web ranking, *PageRank* has become a major method for
evaluating popularity of nodes in information networks. Besides its primary application in search engines, PageRank is successfully used for solving other important problems such as spam detection [1], graph partitioning [2], and finding
gems in scientific citations [3], just to name a few. The PageRank [4] is defined
as a stationary distribution of a random walk on a set of Web pages. At each
step, with probability c, the random walk follows a randomly chosen outgoing
link, and with probability $1-c$, the walk starts afresh from a page chosen at random according to some distribution f. Such random jump also occurs if a page
is *dangling*, i.e. it does not have outgoing links. In the original definition, the
teleportation distribution f is uniform over all Web pages. Then the PageRank
values satisfy the equation

$$PR(i) = c \sum_{j \to i} \frac{1}{d_j} PR(j) + \frac{c}{n} \sum_{j \in \mathcal{D}} PR(j) + \frac{1-c}{n}, \ i = 1, \dots, n, \qquad (1)$$

where $PR(i)$ is the PageRank of page i, d_j is the number of outgoing links of
page j, the sum is taken over all pages j that link to page i, $\mathcal{D}$ is a set of dangling
nodes, and c is the damping factor, which is a constant between 0 and 1.

A. Bonato and F.R.K. Chung (Eds.): WAW 2007, LNCS 4863, pp. 108–123, 2007.
© Springer-Verlag Berlin Heidelberg 2007

From (1) it can be expected that the distribution of PageRank should be related to the distribution of *in-degree*, the number of incoming links. Most of experimental studies of the Web agree that in-degree follows a power law with exponent $\alpha = 1.1$ for cumulative plot. Further research [5,6,7,8] confirmed that PageRank also follows a power law with the same exponent. Mathematical justifications have been proposed in [9,10] for the preferential attachment models [11], and in [12], where the relation between PageRank and in-degree is modelled through a stochastic equation.

At this point, it is important to realize that PageRank is a *global* characteristic of the Web, which depends on in-degrees, *out-degrees*, correlations, and other characteristics of the underlying graph. In contrast to in-degrees, whose impact on the PageRank log-log plot is thoroughly explored and relatively well understood, the influence of out-degrees and dangling nodes has hardly received any attention in the literature. It is however a common belief that dangling nodes are important [13] whereas out-degrees (almost) do not affect the PageRank [7]. We also note that in the literature, there is no common agreement on the out-degree distribution. On the Web data, Broder et al. [14] report a power law with exponent about 2.6 for the density, whereas e.g. Donato et al. [6] obtain a distribution, which is clearly not a power law. On the other hand, for Wikipedia [15], out-degree seems to follow a power law with the same exponent as in-degree.

In the present paper we investigate the relations between PageRank and in/out-degrees, both analytically and experimentally. Our analytical model is an extension of [12]. We view the PageRank of a random page as a random variable R that depends on other factors through a stochastic equation resembling (1). We are concerned with the *tail* probability $\mathbb{P}(R > x)$, i.e. the fraction of pages with PageRank greater than x, when x is large. Our goal is to determine the asymptotic behavior of $\mathbb{P}(R > x)$, that is, we want to find a known function $r(x)$ such that $\mathbb{P}(R > x)/r(x) \to 1$ as $x \to \infty$. In this case, we say that $\mathbb{P}(R > x)$ and $r(x)$ are asymptotically equivalent, which essentially means that for large enough x, $\mathbb{P}(R > x)$ and $r(x)$ are close, and their log-log plots look the same. We formally describe power laws in terms of regular varying random variables, and we use recent results on regular variation to obtain the PageRank asymptotics. To this end, we provide a recurrent stochastic model for the power iteration algorithm commonly used in PageRank computations [16], and we obtain the PageRank asymptotics after each iteration. The necessary background on regular variation can be found in Appendix A.1.

The analytical results suggest that the PageRank and in-degree follow power laws with the same exponent. The out-degrees and dangling nodes affect only a multiplicative constant, for which we find an exact expression. Moreover, it turns out that the out-degree sequence has a truly minor influence. We verify our analytical results in Section 5, where we present the experiments on the Indochina-2004 and EU-2005 Web samples [17], on Wikipedia, and on a synthetic graph. Finally, in Section 6 we suggest a ranking scheme that does not depend on the out-degrees.

2   The Model

2.1   In-Degree

It is a common knowledge that in-degrees in the Web graph obey a power law with exponent about 1.1 for cumulative plot. The power law exponent may deviate somewhat depending on a data set [18] and an estimator [19]. As in our previous work [12], we model the in-degree as an integer regularly varying random variable. To this end, we assume that the in-degree of a random page is distributed as $N(T)$, where T is regularly varying with index α and $N(t)$ is the number of Poisson arrivals on the time interval $[0, t]$. If T is regularly varying then $N(T)$ is also regularly varying and asymptotically identical to T (see e.g. [12]). Thus, $N(T)$ is indeed integer and obeys the power law. To simplify the notation, we will use N instead of $N(T)$ throughout the paper. The proposed formalization for the in-degree distribution allows us to model the number of terms in the summation in (1).

2.2   Out-Degree and Inspection Paradox

Now, we want to model the weights $1/d_j$ in (1). Recall that d_j is the out-degree of page j that has a link to page i. In [12] we studied the relation between in-degree and PageRank assuming that out-degrees of all pages are constant, equal to the expected in-degree d. In this work, we make a step further allowing for random out-degrees.

We model out-degrees of pages linking to a randomly chosen page as independent and identically distributed random variables with arbitrary distribution. Thus, consider a random variable D, which represents the out-degree of a page that links to a particular randomly chosen page i. Note that D is *not* the same random variable as an out-degree of a random page since the additional information that a page has a link to i, alters the out-degree distribution. This famous phenomenon, called *inspection paradox*, finds its mathematical explanations in Renewal Theory. The inspection paradox roughly states that an interval containing a random point tends to be larger than a randomly chosen interval [20]. For instance, in [21], a number of children in a family, to which a randomly chosen child belongs, is stochastically larger than a number of children in a randomly chosen family. Likewise, a number of out-links D from a page containing a random link, should be stochastically larger than an out-degree of a random page. We will refer to D as *effective out-degree*. The term is motivated by the fact that the distribution of D is the one that participates in the PageRank formula.

Now, let p_j be a fraction of pages with out-degree $j \geq 0$. Then we have

$$\lim_{n \to \infty} \mathbb{P}(D = j) = jp_j/d, \quad j \geq 1. \tag{2}$$

where d is the average in/out-degree, and n is the number of pages in the Web. For sufficiently large networks, we may assume that the distribution of D equals to its limiting distribution defined by (2). Note that, naturally, the probability that a random link comes from a page with out-degree k is proportional to k. This was implicitly observed by Fortunato et al. in [7], who in fact used (2) in their computations for the mean-field approximation of the PageRank.

2.3   Stochastic Equation

It is clear that the PageRank values in (1) scale as $1/n$ with the number of pages. In the analysis, it is more convenient to deal with corresponding *scale-free* PageRank scores

$$R(i) = nPR(i), \quad i = 1, \ldots, n, \tag{3}$$

assuming that n goes to infinity. In this setting, it is easier to compare the probabilistic properties of PageRank and in/out-degrees, which are also scale-free. In the remainder of the paper, by PageRank we mean the scale-free PageRank scores (3).

We view the scale-free PageRank of a random page as a random variable R with $\mathbb{E}(R) = 1$. Further, we assume that the PageRank of a random page does not depend on the fact whether the page is dangling. Indeed, it can be shown that the PageRank of a page can not be altered significantly by modifying outgoing links [22]. Moreover, experiments e.g. in [13] show that dangling nodes are often just regular pages whose links have not been crawled. Besides, even authentically dangling pages such as .pdf or .ps files, often contain important information and gain a high ranking independently of the fact that they do not have outgoing links. We note that such independence immediately implies that in large networks, the fraction of the total PageRank mass concentrated in dangling nodes, equals to the fraction of dangling nodes p_0, simply by the law of large numbers: $p_0 = (1/n) \sum_{j \in \mathcal{D}} R(j)$.

Our goal is to model and analyze to which extent the tail probability $\mathbb{P}(R > x)$ for large enough x depends on the in-degree N, the effective out-degree D, and the fraction of dangling nodes p_0. To this end, we model PageRank R as a solution of a stochastic equation involving N and D. Inspired by the original formula (1), the stochastic equation for the scale-free PageRank is as follows:

$$R \stackrel{d}{=} c \sum_{j=1}^{N} \frac{1}{D_j} R_j + [1 - c(1 - p_0)]. \tag{4}$$

Here N, R_j's and D_j's are independent; R_j's are distributed as R, D_j's are distributed as D, and $a \stackrel{d}{=} b$ means that a and b have the same probability distribution. As before, $c \in (0, 1)$ is a damping factor.

We note that the independence assumption for PageRanks and effective out-degrees of pages linking to the same page, is obviously not true in general. However, there is also no direct relation between these values as there is no experimental evidence that such dependencies would crucially influence the PageRank distribution. Thus, we assume independence in this study.

The stochastic equation (4) is a generalization of the equation analyzed in [12], where it was assumed that D_j's are constant. In order to demonstrate applicability of our model, we will use (4) to derive a mean-field approximation for the PageRank of a page with given in-degree. It follows from (2) that

$$\mathbb{E}\left(\frac{1}{D}\right) = \sum_{k=1}^{\infty} \frac{1}{k} \mathbb{P}(D = k) = \sum_{k=1}^{\infty} \frac{1}{k} \frac{kp_k}{d} = \frac{1 - p_0}{d}. \tag{5}$$

Then, assuming that $\mathbb{E}(R_j) = 1$, $j = 1, 2, \ldots$, we obtain

$$\mathbb{E}(R|N) = \frac{c(1 - p_0)}{d} N + [1 - c(1 - p_0)]. \tag{6}$$

If $p_0 = 0$ then this coincides with the mean-field approximation by Fortunato et al. in [7], obtained directly from the PageRank definition under minimal independence assumptions and without considering dangling nodes.

Equation (4) belongs to the class of stochastic recursive equations that were discussed in detail in the recent survey by Aldous and Bandyopadhyay [23]. In particular, (4) has an apparent similarity with distributional equations motivated by branching processes and branching random walks. Such equations were studied in detail by Liu in [24] and his other papers. Taking expectations in (6), we see that if $\mathbb{E}(R_j) = 1$, $j = 1, 2, \ldots$, then $\mathbb{E}(R)$ also equals 1. In Section 4 we will show that (4) has a unique solution R such that $\mathbb{E}(R) = 1$.

3   Power Iterations

In this section, we shall introduce an iteration procedure for solving (4). This procedure can be seen as a stochastic model for the power iteration method commonly used in PageRank computations. We first present the notations, which are in lines with Liu [24].

Let $\{(N_u, 1/D_{u_1}, 1/D_{u_2}, \ldots)\}_u$ be a family of independent copies of $(N, 1/D_1, 1/D_2, \ldots)$ indexed by all finite sequences $u = u_1 \ldots u_n$, $u_i \in \{1, 2, \ldots\}$. And let $\mathbb{T}$ be the Galton-Watson tree with defining elements $\{N_u\}$: we have $\emptyset \in \mathbb{T}$ and, if $u \in \mathbb{T}$ and $i \in \{1, 2, \ldots\}$, then concatenation $ui \in \mathbb{T}$ if and only if $1 \le i \le N_u$. In other words, we indexed the nodes of the tree with root $\emptyset$ and the first level nodes $1, 2, ..N_\emptyset$, and at every subsequent level, the ith offspring of u is named ui (see Figure 1).

Now, we will iterate the equation (4). We start with initial distribution $R^{(0)}$, and for every $k \ge 1$, we define the result of the kth iteration through a distributional identity

$$R^{(k)} \overset{d}{=} c \sum_{j=1}^{N} \frac{1}{D_j} R_j^{(k-1)} + [1 - c(1 - p_0)], \tag{7}$$

where N, $R_j^{(k-1)}$ and D_j, $j \ge 1$, are independent. If $R^{(0)} \equiv 1$ then $R^{(k)}$ serves as a stochastic model for the result of the kth power iteration in standard PageRank computations.

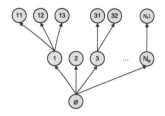

Fig. 1. An example of Galton-Watson tree

Since PageRank vector is always a result of a finite number of iterations, it follows that $R^{(k)}$ describes the distribution of PageRank if the power iteration algorithm stops after k steps. Assuming that in-degrees, effective out-degrees and $R_u^{(0)}$, $u \in \mathbb{T}$, are independent, and repeatedly applying (7), we derive the following representation for $R^{(k)}$, $k \geq 1$:

$$R^{(k)} = c^k \sum_{u=u_1..u_k \in \mathbb{T}} \frac{1}{D_{u_1}} \cdots \frac{1}{D_{u_1..u_k}} R_{u_1..u_k}^{(0)} + [1 - c(1 - p_0)] \sum_{n=0}^{k-1} c^n Y^{(n)} \quad (8)$$

where

$$Y^{(n)} = \sum_{u=u_1...u_n \in \mathbb{T}} \frac{1}{D_{u_1}} \cdots \frac{1}{D_{u_1...u_n}}, \ n \geq 1.$$

The random variable $Y^{(n)}$ represents the sum of the weights of the nth level of the Galton-Watson tree, where the root has weight 1, each edge has a random weight distributed as $1/D$, and the weight of a node is a product of weights of the edges that lead to this node from the root.

In the subsequent analysis we will prove that iterations $R^{(k)}$, $k \geq 1$, converge to a unique solution of (4), and we will obtain the tail behavior of $R^{(k)}$ for each $k \geq 1$. This will give us the asymptotic behavior of the PageRank vector after an arbitrary number of power iterations.

4   Analytical Results

First, we establish that our main stochastic equation (4) indeed defines a unique distribution R, that can serve as a model for the PageRank of a random page. The result is formally stated in the next theorem (the proof is given in Appendix A.2).

Theorem 1. *Equation (4) has a unique non-trivial solution with mean 1 given by*

$$R^{(\infty)} = \lim_{k \to \infty} R^{(k)} = [1 - c(1 - p_0)] \sum_{n=0}^{\infty} c^n Y^{(n)}. \quad (9)$$

Now we are ready to describe the tail behavior of $R^{(k)}$, $k \geq 1$, which models the PageRank after k power iterations. The main result is presented in Theorem 2 below.

Theorem 2. *If* $\mathbb{P}\left(R^{(0)} > x\right) = o(\mathbb{P}(N > x))$, *then for all* $k \geq 1$,

$$\mathbb{P}(R^{(k)} > x) \sim C_k \mathbb{P}(N > x) \text{ as } x \to \infty,$$

where $C_k = \left(\frac{c(1-p_0)}{d}\right)^\alpha \sum_{j=0}^{k-1} c^{j\alpha} b^j$, *and* $b = d\mathbb{E}\left(1/D^\alpha\right) = \sum_{j=1}^{\infty} \frac{p_j}{j^{\alpha-1}}$.

The form of the coefficient C_k arises from the proof (see A.2), which relies on the results from [25]. For large enough k, C_k can be approximated by

$$C = \lim_{k \to \infty} C_k = \frac{c^\alpha (1 - p_0)^\alpha}{d^\alpha (1 - c^\alpha b)}.$$

From the Jensen's inequality $\mathbb{E}(1/D^\alpha) \geq (\mathbb{E}(1/D))^\alpha$ and (5), it follows that $b \geq (1 - p_0)^\alpha d^{1-\alpha}$, and hence,

$$C \geq \frac{c^\alpha (1 - p_0)^\alpha}{d^\alpha (1 - c^\alpha (1 - p_0)^\alpha d^{1-\alpha})}. \tag{10}$$

The last expression is the value of C if out-degree of all non-dangling nodes is a constant. Note that if $\alpha \approx 1.1$, then the difference between the left- and the right-hand sides of (10) is really small for any reasonable out-degree distribution.

From Theorem 2 we see that the power law exponent of the PageRank is the same as the power law exponent of in-degree. Thus, in-degree remains a major factor shaping the PageRank distribution. The multiple factor C_k, $k \geq 1$, depends mainly on the mean in-degree d, damping factor c, and the fraction of non-dangling nodes $(1 - p_0)$. The out-degree distribution $\{p_j, j \geq 1\}$ influences the coefficient b, but this results in a truly minor impact on the PageRank asymptotics. In the next section we will compare out analytical findings with experimental results.

5   Experiments

5.1   Web Data

We performed experiments on Indochina-2004 and EU-2005 Web samples collected by The Laboratory for Web Algorithmics (LAW), Dipartimento di Scienze dell'Informazione (DSI) of the Università degli studi di Milano [17]. In Figures 2,3 below we present cumulative log-log plots for in-degree/PageRank. The y-axis corresponds to the fraction of pages with in-degree/PageRank greater than the value on the x-axis. For in-degree, the power law exponent in evaluated using the maximum likelihood estimator from [19], and the straight line is fitted accordingly. For the PageRank, we plot the *theoretically predicted* straight lines obtained from Theorem 2.

The Indochina set contains 7,414,866 nodes and 194,109,311 links. The results are presented in Figure 2 below. The in-degree plot resembles a power law except for the excessively large fraction of pages with in-degree about 10^4. The presence of *bump* was observed also in other data samples in the past [14,26]. In [26], the authors suggested that it could be probably due to a huge clique created by a single spammer. For more detail on this data set see [18]. For Indochina, we obtain a power law exponent 1.17 for cumulative plot, which is quite different from the result in [18]. This demonstrates the sensitivity of estimators for the power law exponent. Indeed, the exponent 0.6 in [18] reflects the behavior in the first part of the plot, whereas 1.17 gives more weight on the tail of the in-degree distribution.

We fit the straight line $y = -1.17x + 0.8$ into the in-degree plot and then compute the distance

$$\log_{10}(C) = \log_{10}\left(\frac{c^{\alpha}(1 - p_0)^{\alpha}}{d^{\alpha}(1 - c^{\alpha}b)}\right)$$

between the in-degree and the PageRank log-log plots for $c = 0.2, 0.5$, and 0.85. With $d = 26.17$, $p_0 = 0.18$, and $b = 0.65$, we obtain the following prediction for the PageRank log-log plot: $y = -1.17x - 1.73$ for $c = 0.2$, $y = -1.17x - 1.16$ for $c = 0.5$, and $y = -1.17x - 0.70$ for $c = 0.85$. In Figure 2 we show these *theoretically predicted* lines and the experimental PageRank log-log plots. We see that for this data set, our model provides the linear fit with a striking accuracy.

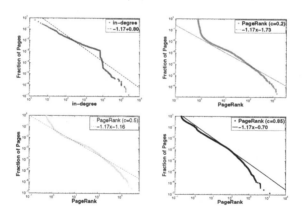

Fig. 2. Indochina data set: cumulative log-log plots for in-degree/PageRank. The straight lines for the PageRank plots are predicted by the model.

We performed the same experiment for EU-2005 of 862,664 nodes and 19,235,140 links. In this data set in-degree shows a typical power law behavior, which is fitted perfectly by $y = -1.1x + 0.61$. We use the same approach to calculate the difference between the in-degree and PageRank plots for $d = 22.3$, $p_0 = 0.08$, $b = 0.70$. Thus, the theoretical prediction for the PageRank are $y = -1.1x - 1.63$, $y = -1.1x - 1.07$, and $y = -1.1x - 0.60$ for $c = 0.2, 0.5$, and 0.85, respectively. The log-log plots for experimental data, the fitted straight line for in-degree, and corresponding theoretical straight lines for PageRank, are presented in Figure 3.

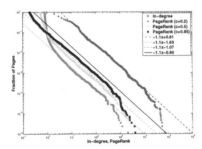

Fig. 3. EU-2005 data set: cumulative log-log plots for in-degree/PageRank. The straight lines for the PageRank plots are predicted by the model.

5.2 Wikipedia

In order to further verify our results, we performed the experiments on the Wikipedia (English) data, whose structure is slightly different from the Web graph [15], in particular, in its out-degree distribution. In Figure 4 we plot the in/out-degree distributions for our data set, which contains 4,881,983 nodes and 42,062,836 links. In Figure 5 we show the in-degree and PageRank plots, with fitted straight line $y = -1.18x + 0.30$ for the in-degree and predicted lines for the PageRank. In this data set we have $d = 8.6159$, $p_0 = 0$ and $b = 0.8668$. Figure 5 (left) shows the PageRank plot for $c = 0.5$ and $c = 0.85$. In Figure 5 (right) we depict the PageRank plots after the first, the second and the last iterations for $c = 0.85$.

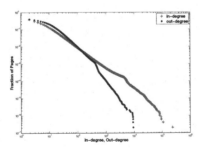

Fig. 4. Wikipedia data set: cumulative log-log plots for in/out-degree

Although the obtained lines do not match perfectly the PageRank plots, we see that our model correctly captures the dynamics of the PageRank distribution in successive power iterations and for different values of c. Most importantly, we observe that the PageRank for Wikipedia retains its power law distribution, and the exponent is again the same as the one for in-degree. Clearly, this crucial property is not influenced by the out-degree distribution.

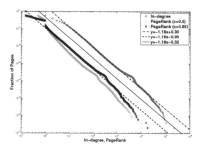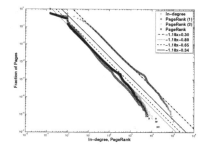

Fig. 5. Wikipedia data set: cumulative log-log plots for PageRank/in-degree and predicted straight lines for different values of c (left) and for different number of iterations (right)

5.3 Synthetic Graph

Next, we performed the experiments on a synthetic graph with out-degree close to constant. The graph of 5,000,000 nodes and 41,577,523 links was generated using the growing network model from [11]. Further, 30% of the links were redirected in order to make the graph more realistic and comparable to Wikipedia. The original out-degree was 9, however, due to the duplicated edges, the average out-degree became 8.3155.

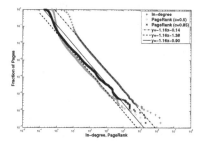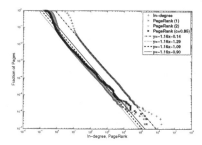

Fig. 6. Synthetic data: cumulative log-log plots for PageRank/in-degree and predicted straight lines for different values of c (left) and for different number of iterations (right)

The results on the PageRank distribution are presented in Figure 6. For the in-degree, we computed $\alpha = 1.14$. The predicted lines for the PageRank are obtained with $d = 8.3155$, $b = 0.7134$ ($\approx d^{1-\alpha}$), $p_0 = 0$. In Figure 6 (left) we show the PageRank plot for $c = 0.5$ and $c = 0.85$, and Figure 6 (right) displays the PageRank plots after the first, the second and the last iterations for $c = 0.85$. One can see that our model provides a good estimation for the difference between the graphs. Furthermore, the lines look parallel as before, although in the growing network models, the PageRank power law exponent is proved to depend on the damping factor [9]. Here we clearly face the fact that the nuances of the 'real' slope are hard to capture on the data. Consequently, our model works well in this case. We note however that in our previous work [12] we studied another version of the growing network graph that showed a quite different behavior.

6 PAR Ranking Scheme

The negligible effect of out-degree distribution on the PageRank behavior made us wonder about the role of out-degrees in link-based ranking in general. In HITS [27], the ranking of a page i is determined by its authority score, which in turn depends on hub scores of pages linking to i. Furthermore, a hub score is high for pages with high out-degree, and thus getting a link from such a page is *advantageous* in HITS whereas it is *disadvantageous* in PageRank according to (1). Since both HITS and PageRank work well in practice, one may try to think of some ranking scheme where out-degree *does not play a role* at all.

We propose one such ranking scheme that we call a Pure Authority Rank (PAR). This algorithm is a mixture between HITS and PageRank. The PAR is defined iteratively. The initial score of each page $i = 1, \ldots, n$ is $s_i^{(0)} = 1/n$, and the results of successive iterations are computed as

$$s_i^{(k)} = \frac{c}{d} \sum_{j \to i} s_j^{(k-1)} + \frac{1-c}{n}, \quad k \geq 1, \tag{11}$$

and then normalized so that $\sum_{i=1}^n s_i^{(k)} = 1$. Here again d is the average in/out-degree and the summation is over all pages j that link to i.

Now, let A be an adjacency matrix of the Web Graph. Then denoting $\tilde{A} = (c/d)A + (1 - c)E/n$, where E is the matrix of ones, we can write (11) with the subsequent normalization in a matrix-vector form as $\mathbf{s}^{(k)} = \mathbf{s}^{(k-1)} \tilde{A} / \|\mathbf{s}^{(k-1)} \tilde{A}\|$, where $\mathbf{s}^{(k)} = (s_1^{(k)}, \ldots, s_n^{(k)})$ and $\| \cdot \|$ is the L_1 norm. Since $\tilde{A}$ is a positive matrix, the convergence and uniqueness of the PAR scores are guaranteed by the Perron-Frobenius theorems. If we take $c = 1$ we obtain an algorithm close to HITS but without the hub-iteration. In this case, the algorithm will converge but the resulting vector might depend on the initial vector, as in HITS and SALSA [28]. We refer to [29] for the detailed uniqueness analysis of link-based ranking schemes.

We computed the PAR scores for Wikipedia and the synthetic graph. The algorithm converges fast and, remarkably, the speed of convergence does not depend on c (see Table 1). In Figure 7 we present the log-log plots for PAR and PageRank. Since the two methods are similar, it is not surprising that the PAR distribution seems to follow a power law with the same exponent as in-degree. A more interesting observation is that the PAR plot for Wikipedia in Figure 7 (right) behaves similar to a PageRank plot computed for a higher value of c.

Finally, we computed Kendall's tau, Spearman's rho, and the correlation coefficient between PAR and PageRank scores for the top 1% pages. The results are presented in Table 1.

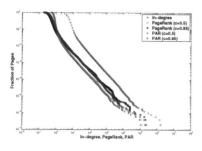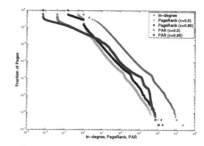

Fig. 7. PageRank and PAR log-log plots: synthetic data (left) and Wikipedia (right)

Table 1. Comparison of PageRank and PAR

Data	c	Scores	Ranks		Iterations	
		Correlation coefficient	Kendall's τ	Spearman's ρ	PR	PAR
Synthetic graph	0.5	0.8112	0.1234	0.1827	8	7
	0.85	0.9753	0.1002	0.1488	13	9
Wikipedia	0.5	0.2474	0.3510	0.4304	8	17
	0.85	0.4675	0.3629	0.4422	29	18

The high correlation between the scores for synthetic graph is expected since in this case the difference between PAR and PageRank is minimal. On the other hand, the correlation between the ranks is on a lower side. Here we observe that the ranking order is very sensitive to the nuances of the algorithm. For a more fair comparison of the two algorithms, future research should reveal which pages were demoted and which were promoted. We believe that the advantages of the PAR algorithm, such as fast convergence and insensitivity to out-degrees, should definitely attract more studies.

Acknowledgments

This work is supported by NWO Meervoud grant no. 632.002.401.

References

1. Gyongyi, Z., Garcia-Molina, H., Pedersen, J.: Combating Web spam with TrustRank. In: VLDB 2004: 30th International Conference on Very Large Data Bases, pp. 576–587 (2004)
2. Andersen, R., Chung, F., Lang, K.: Local graph partitioning using PageRank vectors. In: FOCS 2006: Proceedings of the 47th Annual IEEE Symposium on Foundations of Computer Science, pp. 475–486 (2006)
3. Chen, P., Xie, H., Maslov, S., Redner, S.: Finding scientific gems with Google. Technical Report 0604130, arxiv/physics (2006)
4. Brin, S., Page, L.: The anatomy of a large-scale hypertextual Web search engine. Comput. Networks 33, 107–117 (1998)

5. Pandurangan, G., Raghavan, P., Upfal, E.: Using PageRank to characterize Web structure. In: Ibarra, O.H., Zhang, L. (eds.) COCOON 2002. LNCS, vol. 2387, Springer, Heidelberg (2002)
6. Donato, D., Laura, L., Leonardi, S., Millozi, S.: Large scale properties of the Webgraph. Eur. Phys. J. 38, 239–243 (2004)
7. Fortunato, S., Boguna, M., Flammini, A., Menczer, F.: How to make the top ten: Approximating PageRank from in-degree. Technical Report 0511016, arXiv/cs (2005)
8. Becchetti, L., Castillo, C.: The distribution of PageRank follows a power-law only for particular values of the damping factor. In: WWW 2006: Proceedings of the 15th International Conference on World Wide Web, pp. 941–942. ACM Press, New York (2006)
9. Avrachenkov, K., Lebedev, D.: PageRank of scale-free growing networks. Internet Math. 3(2), 207–231 (2006)
10. Fortunato, S., Flammini, A.: Random walks on directed networks: the case of PageRank. Technical Report 0604203, arXiv/physics (2006)
11. Albert, R., Barabási, A.L.: Emergence of scaling in random networks. Science 286, 509–512 (1999)
12. Litvak, N., Scheinhardt, W.R.W., Volkovich, Y.: In-degree and PageRank of Web pages: Why do they follow similar power laws? Memorandum 1807, University of Twente, Enschede (2006) (to appear in Internet Mathematics)
13. Eiron, N., McCurley, K.S., Tomlin, J.A.: Ranking the Web frontier. In: WWW 2004: Proceedings of the 13th International Conference on World Wide Web, pp. 309–318. ACM Press, New York (2004)
14. Broder, A., Kumar, R., Maghoul, F., Raghavan, P., Rajagopalan, S., Statac, R., Tomkins, A., Wiener, J.: Graph structure in the Web. Comput. Networks 33, 309–320 (2000)
15. Capocci, A., Servedio, V.D.P., Colaiori, F., Buriol, L.S., Donato, D., Leonardiand, S., Caldarelli, G.: Preferential attachment in the growth of social networks: the case of Wikipedia. Technical Report 0602026, arXiv/physics (2006)
16. Langville, A.N., Meyer, C.D.: Deeper inside PageRank. Internet Math. 1, 335–380 (2003)
17. http://law.dsi.unimi.it/ (Accessed in January 2007)
18. Baeza-Yates, R., Castillo, C., Efthimiadis, E.: Characterization of national Web domains. ACM TOIT 7(2) (May 2007)
19. Newman, M.E.J.: Power laws, Pareto distributions and Zipf's law. Cont. Phys. 46, 323–351 (2005)
20. Ross, S.M.: Stochastic processes, 2nd edn. John Wiley & Sons Inc., New York (1996)
21. Ross, S.M.: The inspection paradox. Probab. Engrg. Inform. Sci. 17, 47–51 (2003)
22. Avrachenkov, K., Litvak, N.: The effect of new links on Google PageRank. Stoch. Models 22(2), 319–331 (2006)
23. Aldous, D.J., Bandyopadhyay, A.: A survey of max-type recursive distributional equations. Ann. Appl. Probab. 15, 1047–1110 (2005)
24. Liu, Q.: Asymptotic properties and absolute continuity of laws stable by random weighted mean. Stochastic Process. Appl. 95(1), 83–107 (2001)
25. Jessen, A.H., Mikosch, T.: Regularly varying functions. Publications de l'institut mathematique, Nouvelle série 79(93) (2006)
26. Donato, D., Laura, L., Leonardi, S., Millozzi, S.: The Web as a graph: How far we are. ACM Trans. Inter. Tech. 7(1) (February 2007)

27. Kleinberg, J.M.: Authoritative sources in a hyperlinked environment. JACM 46(5), 604–632 (1999)
28. Lempel, R., Moran, S.: The stochastic approach for link-structure analysis (SALSA) and the TKC effect. Comput. Networks 33(1-6), 387–401 (2000)
29. Farahat, A., LoFaro, T., Miller, J.C., Rae, G., Ward, L.A.: Authority rankings from HITS, PageRank, and SALSA: Existence, uniqueness, and effect of initialization. SISC 27(4), 1181–1201 (2006)
30. Bingham, N.H., Goldie, C.M., Teugels, J.L.: Regular Variation. Cambridge University Press, Cambridge (1989)

Appendix

A.1 Preliminaries on Regular Variation

The theory of regular variation is a natural formalization of power laws. More comprehensive details could be found, for instance, in [30]. We also refer to Jessen and Mikosch [25] for an excellent recent review.

Definition 1. *A function $L(x)$ is* slowly varying *if for every $t > 0$,*

$$\frac{L(tx)}{L(x)} \to 1 \quad as \quad x \to \infty.$$

Definition 2. *A* non-negative random variable X is said to be *regularly varying with index α if*

$$\mathbb{P}(X > x) \sim x^{-\alpha}L(x) \quad as \quad x \to \infty, \tag{A.1}$$

for some positive slowly varying function $L(x)$.

Here, as in the remainder of this paper, the notation $a(x) \sim b(x)$ means that $a(x)/b(x) \to 1$.

The *asymptotic* equivalence (A.1) is a formalization of a power law. In words, it means that for large enough x, the tail distribution $\mathbb{P}(X > x)$ can be approximated by the regularly varying function $x^{-\alpha}L(x)$, which is, in turn, approximately proportional to $x^{-\alpha}$ due to the definition of L.

Regularly varying random variables represent a subclass of a much broader class of long-tailed random variables.

Definition 3. *A* random variable X is long-tailed *if for any $y > 0$,*

$$\mathbb{P}(X > x + y) \sim \mathbb{P}(X > x) \quad as \quad x \to \infty. \tag{A.2}$$

Next lemma describes the behavior of a product and random sums of regular varying random variables. The relation (i) is known as Breiman's theorem (see e.g. Lemma 4.2.(1) in [25]). Properties (ii) and (iii) are, respectively, statements (2) and (5) of Lemma 3.7 in [25].

Lemma A.1. (i) *Assume that X_1 and X_2 are two independent non-negative random variables such that X_1 is regularly varying with index α and that $\mathbb{E}(X_2^{\alpha+\epsilon}) < \infty$ for some $\epsilon > 0$. Then*

$$\mathbb{P}(X_1 X_2 > x) \sim \mathbb{E}(X_2^{\alpha})\mathbb{P}(X_1 > x).$$

(ii) *Assume that N is regularly varying with index $\alpha \geq 0$; if $\alpha = 1$, then assume that $\mathbb{E}(N) < \infty$. Moreover, let (X_i) be i.i.d. sequence such that $\mathbb{E}(X_1) < \infty$ and $\mathbb{P}(X_1 > x) = o(\mathbb{P}(N > x))$. Then as $x \to \infty$,*

$$P(\sum_{i=1}^{N} X_i > x) \sim (\mathbb{E}(X_1))^{\alpha} P(N > x).$$

(iii) *Assume that $P(N > x) \sim r P(X_1 > x)$ for some $r > 0$, that X_1 is regularly varying with index $\alpha \geq 1$, and $E(X_1) < \infty$. Then*

$$P(\sum_{i=1}^{N} X_i > x) \sim (\mathbb{E}(N) + r(\mathbb{E}(X_1))^{\alpha})P(X_1 > x).$$

A.2   Proofs

Proof (of Theorem 1)
It is easy to verify that $R^{(\infty)}$ in (9) is a well-defined solution of (4). In particular, according to the monotone convergence theorem,

$$\mathbb{E}(R^{(\infty)}) = [1 - c(1 - p_0)] \lim_{k \to \infty} \sum_{n=1}^{k} c^n \mathbb{E}(Y^{(n)}) = 1.$$

To prove the uniqueness, assume that there is another solution with mean 1 and take this solution as an initial distribution $R^{(0)}$ with $\mathbb{E}(R^{(0)}) = 1$. Now, the first part of (8) has a mean $c^k(1 - p_0)^k$, and hence it converges in probability to 0 because, by the Markov inequality, the probability that this term is greater than some $\epsilon > 0$ is at most $c^k(1 - p_0)^k/\epsilon \to 0$ as $k \to \infty$. Moreover, the second part of (8) converges a.s. to $R^{(\infty)}$ as $k \to \infty$. It follows that (8) converges to $R^{(\infty)}$ in probability. We conclude that there is no other fixed point of (4) with mean 1 except $R^{(\infty)}$.

Proof (of Theorem 2). We will use the induction. For $k = 1$, we derive

$$\mathbb{P}\left(R^{(1)} > x\right) \sim \mathbb{P}\left(\sum_{j=1}^{N} \frac{c}{D_j} R_j^{(0)} + [1 - c(1 - p_0)] > x\right)$$

$$\sim \left(\frac{c(1 - p_0)}{d}\right)^{\alpha} \mathbb{P}(N > x - [1 - c(1 - p_0)])$$

$$\sim C_1 \mathbb{P}(N > x) \text{ as } x \to \infty,$$

where the second relation follows from Lemma A.1(ii) because $\mathbb{E}(N) = d < \infty$, $\mathbb{E}\left(R_1^{(0)}\right) = 1$, $\mathbb{E}\left(cD_1^{-1}R_1^{(0)}\right) = c(1-p_0)d^{-1} < \infty$, and $\mathbb{P}\left(cD_1^{-1}R_1^{(0)} > x\right) = o(\mathbb{P}(N > x))$, and the last relation follows from (A.2).

Now, assume that the result has been shown for $(k-1)$th iteration, $k \geq 2$. Then Lemma A.1(i) yields

$$\mathbb{P}\left(\frac{c}{D}R^{(k-1)} > x\right) \sim c^{\alpha}\mathbb{E}\left(\frac{1}{D^{\alpha}}\right)C_{k-1}\mathbb{P}(N > x),$$

$$= \frac{c^{\alpha}}{d}\, b\, C_{k-1}\mathbb{P}(N > x),$$

where

$$\mathbb{E}\left(\frac{1}{D^{\alpha}}\right) = \sum_{j=1}^{\infty}\frac{p_j}{j^{\alpha}} = \frac{1}{d}\sum_{j=1}^{\infty}\frac{p_j}{j^{\alpha-1}} = \frac{1}{d}b.$$

Then, since $\mathbb{E}\left(cD^{-1}R^{(k-1)}\right) = c(1-p_0)d^{-1} < \infty$ and $\mathbb{E}(N) = d$, we apply Lemma A.1(iii) to obtain

$$\mathbb{P}(R^{(k)} > x) \sim \mathbb{P}\left(\sum_{j=1}^{N}\frac{c}{D_j}R^{(k-1)} + [1 - c(1-p_0)] > x\right)$$

$$\sim \left(c^{\alpha}bC_{k-1} + \left(\frac{c(1-p_0)}{d}\right)^{\alpha}\right)\mathbb{P}(N > x - [1 - c(1-p_0)])$$

$$\sim \left(c^{\alpha}bC_{k-1} + \left(\frac{c(1-p_0)}{d}\right)^{\alpha}\right)\mathbb{P}(N > x) \text{ as } x \to \infty,$$

for any $k \geq 2$. Here the last relation again follows from the property of long-tailed random variables (A.2).

Then for the constant C_k we have

$$C_k = c^{\alpha}\, b\, C_{k-1} + \left(\frac{c(1-p_0)}{d}\right)^{\alpha}$$

$$= \left(c^{\alpha}b\left(\frac{c(1-p_0)}{d}\right)^{\alpha}\sum_{j=0}^{k-2}c^{j\alpha}b^j + \left(\frac{c(1-p_0)}{d}\right)^{\alpha}\right)$$

$$= \left(\frac{c(1-p_0)}{d}\right)^{\alpha}\sum_{j=0}^{k-1}c^{j\alpha}b^j.$$

Approximating Betweenness Centrality

David A. Bader, Shiva Kintali, Kamesh Madduri, and Milena Mihail

College of Computing
Georgia Institute of Technology
{bader, kintali, kamesh, mihail}@cc.gatech.edu

Abstract. Betweenness is a centrality measure based on shortest paths, widely used in complex network analysis. It is computationally-expensive to exactly determine betweenness; currently the fastest-known algorithm by Brandes requires $O(nm)$ time for unweighted graphs and $O(nm + n^2 \log n)$ time for weighted graphs, where n is the number of vertices and m is the number of edges in the network. These are also the worst-case time bounds for computing the betweenness score of a single vertex. In this paper, we present a novel approximation algorithm for computing betweenness centrality of a given vertex, for both weighted and unweighted graphs. Our approximation algorithm is based on an adaptive sampling technique that significantly reduces the number of single-source shortest path computations for vertices with high centrality. We conduct an extensive experimental study on real-world graph instances, and observe that our random sampling algorithm gives very good betweenness approximations for biological networks, road networks and web crawls.

1 Introduction

One of the fundamental problems in network analysis is to determine the *importance* (or the *centrality*) of a particular vertex (or an edge) in a network. Some of the well-known metrics for computing centrality are closeness [1], stress [2] and betweenness [3,4]. Of these indices, betweenness has been extensively used in recent years for the analysis of social-interaction networks, as well as other large-scale complex networks. Some applications include lethality in biological networks [5,6,7], study of sexual networks and AIDS [8], identifying key actors in terrorist networks [9,10], organizational behavior [11], and supply chain management processes [12]. Betweenness is also used as the primary routine in popular algorithms for clustering and community identification [13] in real-world networks. For instance, the Girvan-Newman [14] algorithm iteratively partitions a network by identifying edges with high betweenness scores, removing them and recomputing centrality scores.

Betweenness is a global centrality metric that is based on shortest-path enumeration. Consider a graph $G = (V, E)$, where V is the set of vertices representing *actors* or *nodes* in the complex network, and E, the set of edges representing the relationships between the vertices. The number of vertices and edges are denoted by n and m respectively. The graphs can be directed or undirected.

A. Bonato and F.R.K. Chung (Eds.): WAW 2007, LNCS 4863, pp. 124–137, 2007.

We will assume that each edge $e \in E$ has a positive integer weight $w(e)$. For unweighted graphs, we use $w(e) = 1$. A *path* from vertex s to t is defined as a sequence of edges $\langle u_i, u_{i+1} \rangle$, $0 \le i < l$, where $u_0 = s$ and $u_l = t$. The *length* of a path is the sum of the weights of edges. We use $d(s, t)$ to denote the distance between vertices s and t (the minimum length of any path connecting s and t in G). Let us denote the total number of shortest paths between vertices s and t by λ_{st}, and the number passing through vertex v by $\lambda_{st}(v)$. Let $\delta_{st}(v)$ denote the *fraction* of shortest paths between s and t that pass through a particular vertex v i.e., $\delta_{st}(v) = \frac{\lambda_{st}(v)}{\lambda_{st}}$. We call $\delta_{st}(v)$ the *pair-dependency* of s, t on v.

Betweenness centrality [3,4] of a vertex v is defined as

$$BC(v) = \sum_{s \neq v \neq t \in V} \delta_{st}(v)$$

Currently, the fastest known algorithm for exactly computing betweenness of all the vertices, designed by Brandes [15], requires at least $O(nm)$ time for unweighted graphs and $O(nm + n^2 \log n)$ time for weighted graphs, where n is the number of vertices and m is the number of edges. Thus, for large-scale graphs, exact centrality computation on current workstations is not practically viable. In prior work, we explored high performance computing techniques [16] that exploit the typical small-world graph topology to speed up exact centrality computation. We designed novel parallel algorithms to exactly compute various centrality metrics, optimized for real-world networks. We also demonstrate the capability to compute exact betweenness on several large-scale networks (vertices and edges in the order of millions) from the Internet and social interaction data; these networks are three orders of magnitude larger than instances that can be processed by current social network analysis packages.

Fast centrality estimation is thus an important problem, as a good approximation would be an acceptable alternative to exact scores. Currently the fastest exact algorithms for shortest path enumeration-based metrics require n shortest-path computations; however, it is possible to estimate centrality by extrapolating scores from a fewer number of path computations. Using a random sampling technique, Eppstein and Wang [17] show that the closeness centrality of all vertices in a weighted, undirected graph can be approximated with high probability in $O(\frac{\log n}{\epsilon^2}(n \log n + m))$ time, and an additive error of at most $\epsilon \Delta_G$ (ϵ is a fixed constant, and Δ_G is the diameter of the graph). However, betweenness centrality scores are harder to estimate, and the quality of approximation is found to be dependent on the vertices from which the shortest path computations are initiated from (in this paper, we will refer to them as the set of *source vertices* for the approximation algorithm). Recently, Brandes and Pich [18] presented centrality estimation heuristics, where they experimented with different strategies for selecting the source vertices. They observe that a random selection of source vertices is superior to deterministic strategies. In addition to exact parallel algorithms, we also discussed parallel techniques to compute approximate betweenness centrality in [16], using a random source selection strategy.

While prior approaches approximate centrality scores of *all* vertices in the graph, there are no known algorithms to compute the centrality of a single vertex in time faster than computing the betweenness of all vertices. In this paper, we present a novel *adaptive sampling*-based algorithm for approximately computing betweenness centrality of a given vertex. Our primary result is as follows:

Theorem: For $0 < \epsilon < 0.5$, if the centrality of a vertex v is n^2/t for some constant $t \geq 1$, then with probability $\geq 1 - 2\epsilon$ its centrality can be estimated to within a factor of $1/\epsilon$ with ϵt samples of source vertices.

The rest of this paper is organized as follows. We review the currently-known fastest sequential algorithm by Brandes in Section 2. We present our approximation algorithm based on adaptive sampling and its analysis in Section 3. Section 4 is an experimental study of our approximation technique on several real-world networks. We conclude with a summary of open problems in Section 5.

2   Exact Computation of Betweenness Centrality

Brandes' algorithm [15] shows how to compute centrality scores of all the vertices in the graph in the same asymptotic time bounds as n SSSP computations.

2.1   Brandes' Algorithm

Define the *dependency* of a source vertex $s \in V$ on a vertex $v \in V$ as $\delta_{s*}(v) = \sum_{t \neq s \neq v \in V} \delta_{st}(v)$. Then the betweenness score of v can be then expressed as $BC(v) = \sum_{s \neq v \in V} \delta_{s*}(v)$. Also, let $P_s(v)$ denote the set of *predecessors* of a vertex v on shortest paths from s: $P_s(v) = \{u \in V : \langle u, v \rangle \in E, d(s, v) = d(s, u) + w(u, v)\}$. Brandes shows that the dependencies satisfy the following recursive relation, which is the most crucial step in the algorithm analysis.

Theorem 1. *The dependency of $s \in V$ on any $v \in V$ obeys*

$$\delta_{s*}(v) = \sum_{w:v \in P_s(w)} \frac{\lambda_{sv}}{\lambda_{sw}} (1 + \delta_{s*}(w))$$

First, n SSSP computations are done, one for each $s \in V$. The predecessor sets $P_s(v)$ are maintained during these computations. Next, for every $s \in V$, using the information from the shortest paths tree and predecessor sets along the paths, compute the dependencies $\delta_{s*}(v)$ for all other $v \in V$. To compute the centrality value of a vertex v, we finally compute the sum of all dependency values. The $O(n^2)$ space requirements can be reduced to $O(m+n)$ by maintaining a *running centrality score*. For additional details, we refer the reader to Brandes' paper [15].

3 Adaptive-Sampling Based Approximation

The adaptive sampling technique was introduced by Lipton and Naughton [19] for estimating the size of the transitive closure of a digraph. Prior to their work, algorithms for estimating transitive closure were based on randomly sampling source-vertices, solving the single-source reachability problem for the sampled vertices, and using this information to estimate the size of the transitive closure. The Lipton-Naughton algorithm introduces adaptive sampling of source-vertices, that is, the number of samples varies with the information obtained from each sample.

In this section, we give an *adaptive sampling* algorithm for computing betweenness of a given vertex v. It is a sampling algorithm in that it estimates the centrality by sampling a subset of vertices and performing SSSP computations from these vertices. It is termed *adaptive*, because the number of samples required varies with the information obtained from each sample.

The following lemma is easy to see and the proof is omitted.

Lemma 1. *$BC(v)$ is zero iff its neighboring vertices induce a clique.*

Let a_i denote the dependency of the vertex v_i on v i.e., $a_i = \delta_{v_i *}(v)$. Let $A = \sum a_i = BC(v)$. It is easy to verify that $0 \le a_i \le n - 2$ and $0 \le A \le (n - 1)(n - 2)/2$. The quantity we wish to estimate is A. Consider doing so with the following algorithm:

Algorithm 1. *Repeatedly sample a vertex $v_i \in V$; perform SSSP (using BFS or Dijkstra's algorithm) from v_i and maintain a running sum S of the dependency scores $\delta_{v_i *}(v)$. Sample until S is greater than cn for some constant $c \ge 2$. Let the total number of samples be k. The estimated betweenness centrality score of v, $BC(v)$ is given by $\dfrac{nS}{k}$.*

Let X_i be the random variable representing the dependency of a randomly sampled vertex on v. The probability of an event x is denoted by $\Pr[\,x\,]$. We establish the following lemmas to analyze the above algorithm.

Lemma 2. *Let $E[X_i]$ denote the expectation of X_i and $Var[X_i]$ denote the variance of X_i. Then, $E[X_i] = A/n$, $E[X_i{}^2] \le A$, and $Var[X_i] \le A$.*

The next lemma is useful in proving a lower bound on the expected number of samples made before stopping. The proof is presented in the Appendix.

Lemma 3. *Let $k = \epsilon n^2 / A$. Then,*

$$\Pr[\,X_1 + X_2 + \cdots + X_k \ge cn\,] \le \frac{\epsilon}{(c - \epsilon)^2}$$

Lemma 4. *Let $k \ge \epsilon n^2 / A$ and $d > 0$. Then*

$$\Pr\left[\,\left|\frac{n}{k}\left(\sum_{i=1}^{k} X_i\right) - A\right| \ge dA\,\right] \le \frac{1}{\epsilon d^2}$$

Theorem 2. *Let $\tilde{A}$ be the estimate of A in the above procedure and let $A > 0$. Then for $0 < \epsilon < 0.5$ with probability $\geq 1 - 2\epsilon$, **Algorithm 1** estimates A to within a factor of $1/\epsilon$.*

Proof. There are two ways that the algorithm can fail: (i) it can stop too early to guarantee a good error bound, (ii) it can stop after enough samples but with a bad estimate.

First we claim that the procedure is unlikely to stop with $k \leq n^2/A$. We have that

$$\Pr\left[\,(\exists j)(j \leq k) \wedge (X_1 + X_2 + \cdots + X_j \geq cn)\,\right] \leq \Pr\left[\,X_1 + X_2 + \cdots + X_k \geq cn\,\right]$$

where $k = \dfrac{\epsilon n^2}{A}$, because the event to the right of the inequality implies the event to the left. But by Lemma 3, the right side of this equation is at most $\epsilon/(c-\epsilon)^2$. Substituting $c = 2$ and noting that $0 < \epsilon < 0.5$, we get that this probability is less than ϵ.

Next we turn to the accuracy of the estimate. If $k = \epsilon n^2/A$, by Lemma 4 the estimate,

$$\tilde{A} = \frac{n}{k} \sum_{i=1}^{k} X_i$$

is within dA of A with probability $\geq 1/(\epsilon d^2)$. Letting $d = 1/\epsilon$, this is just ϵ.

Putting the two ways of failure together, we get that the total probability of failure is less than $\epsilon + (1 - \epsilon)\epsilon$, which is less than 2ϵ. Finally, note that if $A > 0$, there must be at least one i such that $a_i > 0$, so the algorithm will terminate. The case when $A = 0$ (i.e., centrality of v is 0) can be detected using Lemma 1 (before running the algorithm). $\qquad\blacksquare$

An interesting aspect of our theorem is that the sampling is adaptive. usually such sampling procedures perform a fixed number of samples. Here it is critical that the algorithm adapts it behavior. Substituting $A = \dfrac{n^2}{t}$ in our analysis we get the following theorem.

Theorem 3. *For $0 < \epsilon < 0.5$, if the centrality of a vertex v is n^2/t for some constant $t \geq 1$, then with probability $\geq 1 - 2\epsilon$ its centrality can be estimated to within a factor of $1/\epsilon$ with ϵt samples of source vertices.*

Although our theoretical result is valid only for high centrality nodes, our experimental results (presented in the next section) show a similar behavior for all the vertices.

4   Experimental Study

We assess the quality of the sampling-based approximation algorithm on several real-world graph instances (see Table 1). We use the parallel centrality analysis toolkit SNAP [20] to compute exact betweenness scores. Since the execution time and speedup achieved by the approximation approach are directly proportional to the number of BFS/shortest path computations, we do not report performance results in this section. For a detailed discussion of exact centrality computation in parallel, and optimizations for small-world graphs, please refer to [16].

Network Data

We experiment with two synthetic graph instances and four real networks in this study. `rand` is an unweighted, undirected random network of 2000 vertices and 7980 edges, generated using the Erdős–Rényi graph model [27]. This synthetic graph has a low diameter, low clustering, and a Gaussian degree distribution. `pref-attach` is a synthetic graph generated using the Preferential attachment model proposed by Barabási and Albert [28]. This model generates graphs with heavy-tailed degree distributions and scale-free properties. Vertices are added one at a time, and for each of them, we create a fixed number of edges connecting to existing vertices, with probability proportional to their degree. `bio-pin` is a biological network that represents interactions in the human proteome [29,23]. This graph is undirected, unweighted and exhibits small-world characteristics. `crawl` corresponds to the `wb-cs-stanford` network in the UF sparse matrix collection [24]. It is a directed graph, where vertices correspond to pages in the Stanford Computer Science domain, and edges represent links. `cite` is a directed graph from the Pajek network collection [25]. It corresponds to papers by and citing J. Lederberg (1945-2002). `road` is a weighted graph of 3353 vertices and 4435 edges that corresponds to a large portion of the road network of Rome, Italy from 1999 [26]. Vertices correspond to intersections between roads, and edges correspond to roads or road segments. Edge weights are physical distances in metres. Road networks have more structure and a higher diameter than the other networks considered in this study.

Table 1. Networks used in the experimental study

Label	Network	n	m	Details	Source
rand	random graph	2000	7980	synthetic, undirected	[21]
pref-attach	preferential attachment	2000	7980	synthetic, undirected	[22]
bio-pin	human protein interactions	8503	32,191	undirected	[23]
crawl	web-crawl (`stanford.edu`)	9914	36,854	directed	[24]
cite	Lederberg citation network	8843	41,601	directed	[24,25]
road	Rome, Italy road network	3353	4435	weighted, undirected	[26]

Methodology

Our goal in this study is to quantify the approximation quality, and so we primarily compare the approximation results to exact scores. We first compute exact centrality scores of all the networks in Table 1. In most data sets, we are interested in high-centrality vertices, as they are the critical entities and are used in further analysis. From the exact scores, we identify vertices whose centrality scores are an order of magnitude greater than the rest of the network. For these vertices, we study the trade-off between computation and approximation quality by varying the parameter c in Algorithm 1. We also show that it is easy to estimate scores of low-centrality vertices. We chose small networks for ease of analysis and visualization, but the approximation algorithm can be effectively applied to large networks as well (see, for instance, the networks considered in [16]).

Experiments

Figure 1 plots the distribution of exact and approximate betweenness scores for the six different test instances. Note that the synthetic networks, rand and pref-attach show significantly lower variation in exact centrality scores compared to the real instances. Also, there are a significant percentage of low-centrality vertices (scores less than, or close to, n) in cite, crawl and bio-pin.

We apply Algorithm 1 to estimate betweenness centrality scores of all the vertices in the test instances. In order to visualize the data better, we plot a smoothed curve of the estimated betweenness centrality data that is superimposed with the exact centrality score scatter-plot. We set the parameter c in Algorithm 1 to 5 for these experiments. In addition, we impose a cut-off of $\frac{n}{20}$ on the number of samples. Observe that in all the networks, the estimated centrality scores are very close to the exact ones, and we are guaranteed to cut down on the computation by a factor of nearly 20.

To further study the quality of approximation for high-centrality vertices, we select the top 1% of the vertices (about 30) ordered by exact centrality score in each network, and compute their estimated centrality scores using the adaptive-sampling algorithm. Since the source vertices in the adaptive approach are chosen randomly, we repeat the experiment five times for each vertex and report the mean and variance in approximation error. Figure 2 plots the mean percentage approximation error in the computed scores for these high centrality vertices, when the value of c (see Algorithm 3) is set to 5. The vertices are sorted by exact centrality score on the X-axis. The error bars in the charts indicate the variance in estimated score due to random runs, for each network. For the random graph instance, the average error is about 5%, while it is roughly around 10% for the rest of the networks. Except for a few anomalous vertices, the error variance is within reasonable bounds in all the graph classes.

Figure 3 plots the percentage of BFS/SSSP computations required for approximating the centrality scores, when c is set to 5. This algorithmic count is an indicator of the amount of work done by the approximation algorithm. The vertices are ordered again by their exact centrality scores from left to right, with the vertex with the least score to the left. A common trend we observe across all graph classes is that the percentage of source vertices decreases as the centrality score increases – this implies that the scores of high centrality vertices can be approximated with lesser work using the adaptive sampling approach. Also,

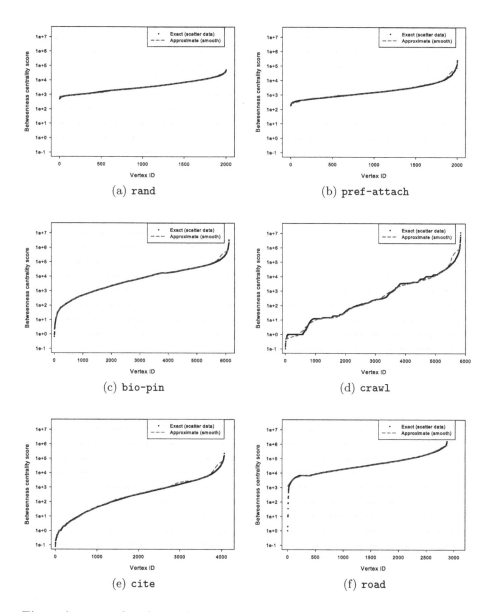

Fig. 1. A scatter plot of exact betweenness scores of all the vertices (in sorted order), and a line plot of their estimated betweenness scores (the approximate betweenness scatter data is smoothed by a local smoothing technique using polynomial regression)

this value is significantly lower for `crawl`, `bio-pin` and `road` compared to other graph classes.

We can also vary the parameter c, which affects both the percentage of BFS/SSSP computations and the approximation quality. Table 2 summarizes

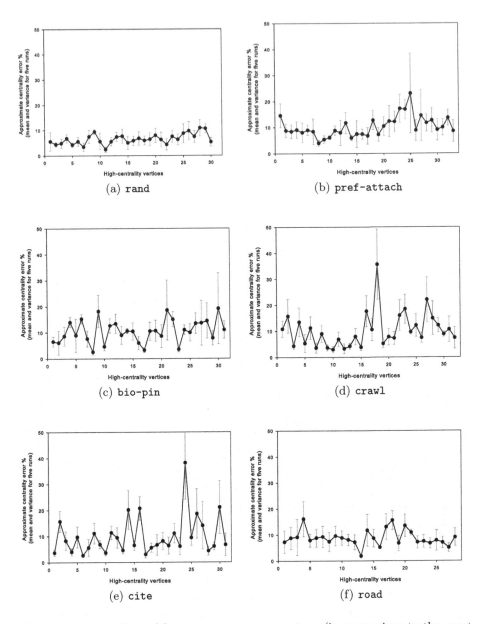

Fig. 2. Average estimated betweenness error percentage (in comparison to the exact centrality score) for multiple runs. The adaptive sampling parameter c is set to 5 for all experiments and the error bars indicate the variance.

the average performance on each graph instance, for different values of c. Taking only high-centrality vertices into consideration, we report the mean approximation error and the number of samples for each graph instance. As expected, we

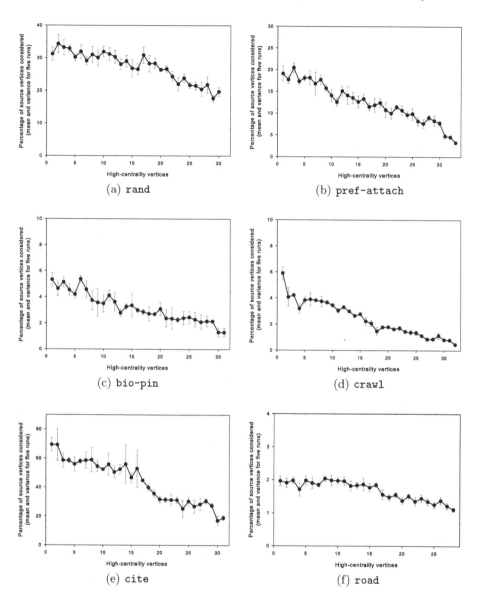

Fig. 3. The number of samples/SSSP computations as a fraction of n, the total number of vertices. This algorithmic count is an indicator of the amount of work done by the approximation algorithm. The adaptive sampling parameter c is set to 5, and the error bars indicate the variance from 5 runs.

find that the error decreases as the parameter c is increased, while the number of samples increases. Since the highest centrality value is around $10 * n$ for the citation network, a significant number of shortest path computations have to be done even for calculating scores with a reasonable accuracy. But for other

Table 2. Observed average-case algorithmic counts, as the value of the sampling parameter c is varied. The average error percentage is the deviation of the estimated score from the exact score, and the $\frac{k}{n}$ percentage indicates the number of samples/SSSP computations.

Network	rand	pref-attach	bio-pin	crawl	cite	road
t = 2						
Avg. error	16.28%	29.39%	46.72%	33.69%	32.51%	22.58%
Avg. $\frac{k}{n}$	11.31%	5.36%	1.30%	0.96%	17.00%	0.68 %
t = 5						
Avg. error	6.51%	10.28%	10.49%	10.31%	9.98%	8.79%
Avg. $\frac{k}{n}$	27.37%	12.38%	3.20%	2.42%	43.85%	1.68%
t = 10						
Avg. error	5.62%	6.13%	7.17%	7.04%	–	7.39%
Avg. $\frac{k}{n}$	54.51%	24.66%	6.33%	4.89%	–	3.29%

graph instances, particularly the road network, web-crawl and the protein interaction network, $c = 5$ offers a good trade-off between computation and the approximation quality.

5   Conclusion and Open Problems

We presented a novel approximation algorithm for computing betweenness centrality, of a given vertex, in both weighted and unweighted graphs. Our approximation algorithm is based on an adaptive sampling technique that significantly reduces the number of single-source shortest path computations for vertices with high centrality. We conduct an extensive experimental study on real-world graph instances, and observe that the approximation algorithm performs well on web crawls, road networks and biological networks.

Approximating the centrality of *all* vertices in time less than $O(nm)$ for unweighted graphs and $O(nm + n^2 \log n)$ for weighted graphs is a challenging open problem. Designing a fully dynamic algorithm for computing betweenness is very useful.

Acknowledgments

The authors are grateful to the ARC (Algorithms and Randomness Center) of the College of Computing, Georgia Institute of Technology, for funding this project. This work was also supported in part by NSF Grants CAREER CCF-0611589, NSF DBI-0420513, ITR EF/BIO 03-31654, and DARPA Contract NBCH 30390004. Kamesh Madduri's work is supported in part by the NASA Graduate Student Researcher Program Fellowship (NASA NP-2005-07-375-HQ). We thank Richard Lipton for helpful discussions.

References

1. Sabidussi, G.: The centrality index of a graph. Psychometrika 31, 581–603 (1966)
2. Shimbel, A.: Structural parameters of communication networks. Bulletin of Mathematical Biophysics 15, 501–507 (1953)
3. Freeman, L.: A set of measures of centrality based on betweenness. Sociometry 40(1), 35–41 (1977)
4. Anthonisse, J.: The rush in a directed graph. In: Report BN9/71, Stichting Mathematisch Centrum, Amsterdam, Netherlands (1971)
5. Jeong, H., Mason, S., Barabási, A.L., Oltvai, Z.: Lethality and centrality in protein networks. Nature 411, 41–42 (2001)
6. Pinney, J., McConkey, G., Westhead, D.: Decomposition of biological networks using betweenness centrality. In: McLysaght, A., Huson, D.H. (eds.) RECOMB 2005. LNCS (LNBI), vol. 3678, Springer, Heidelberg (2005)
7. del Sol, A., Fujihashi, H., O'Meara, P.: Topology of small-world networks of protein-protein complex structures. Bioinformatics 21(8), 1311–1315 (2005)
8. Liljeros, F., Edling, C., Amaral, L., Stanley, H., Åberg, Y.: The web of human sexual contacts. Nature 411, 907–908 (2001)
9. Krebs, V.: Mapping networks of terrorist cells. Connections 24(3), 43–52 (2002)
10. Coffman, T., Greenblatt, S., Marcus, S.: Graph-based technologies for intelligence analysis. Communications of the ACM 47(3), 45–47 (2004)
11. Buckley, N., van Alstyne, M.: Does email make white collar workers more productive? Technical report, University of Michigan (2004)
12. Cisic, D., Kesic, B., Jakomin, L.: Research of the power in the supply chain. In: International Trade, Economics Working Paper Archive EconWPA (April 2000)
13. Newman, M.: The structure and function of complex networks. SIAM Review 45(2), 167–256 (2003)
14. Girvan, M., Newman, M.: Community structure in social and biological networks. Proceedings of the National Academy of Sciences, USA 99(12), 7821–7826 (2002)
15. Brandes, U.: A faster algorithm for betweenness centrality. J. Mathematical Sociology 25(2), 163–177 (2001)
16. Bader, D., Madduri, K.: Parallel algorithms for evaluating centrality indices in real-world networks. In: ICPP. Proc. 35th Int'l Conf. on Parallel Processing, Columbus, OH, IEEE Computer Society, Los Alamitos (2006)
17. Eppstein, D., Wang, J.: Fast approximation of centrality. In: SODA 2001. Proc. 12th Ann. Symp. Discrete Algorithms, Washington, DC, pp. 228–229 (2001)
18. Brandes, U., Pich, C.: Centrality estimation in large networks. In: To appear in Intl. Journal of Bifurcation and Chaos, Special Issue on Complex Networks' Structure and Dynamics (2007)
19. Lipton, R., Naughton, J.: Estimating the size of generalized transitive closures. In: VLDB, pp. 165–171 (1989)
20. Madduri, K., Bader, D.: Small-world Network Analysis in Parallel: a toolkit for centrality analysis (2007), http://www.cc.gatech.edu/~kamesh
21. Madduri, K., Bader, D.: GTgraph: A suite of synthetic graph generators (2006), http://www.cc.gatech.edu/~kamesh/GTgraph
22. Barabási, A.L.: Network databases (2007), http://www.nd.edu/~networks/resources.htm
23. Bader, D., Madduri, K.: A graph-theoretic analysis of the human protein interaction network using multicore parallel algorithms. In: HiCOMB 2007. Proc. 6th Workshop on High Performance Computational Biology, Long Beach, CA (March 2007)

24. Davis, T.: University of Florida Sparse Matrix Collection (2007),
    `http://www.cise.ufl.edu/research/sparse/matrices`
25. Batagelj, V., Mrvar, A.: PAJEK datasets (2006),
    `http://www.vlado.fmf.uni-lj.si/pub/networks/data/`
26. Demetrescu, C., Goldberg, A., Johnson, D.: 9th DIMACS implementation challenge
    – Shortest Paths (2006), `http://www.dis.uniroma1.it/~challenge9/`
27. Erdős, P., Rényi, A.: On random graphs I. Publicationes Mathematicae 6, 290–297
    (1959)
28. Barabási, A.L., Albert, R.: Emergence of scaling in random networks. Science 286(5439), 509–512 (1999)
29. Peri, S., et al.: Development of human protein reference database as an initial
    platform for approaching systems biology in humans. Genome Research 13, 2363–2371 (2003)

Appendix

Lemma 3

Let $k = \epsilon n^2 / A$. Then,

$$\Pr[\, X_1 + X_2 + \cdots + X_k \geq cn \,] \leq \frac{\epsilon}{(c - \epsilon)^2}$$

Proof. We have

$$\Pr[\, X_1 + \cdots + X_k \geq cn \,] = \Pr\left[\, \left(X_1 - \frac{A}{n}\right) + \cdots + \left(X_k - \frac{A}{n}\right) \geq cn - \frac{kA}{n} \,\right]$$

$$= \Pr\left[\, \left(X_1 - \frac{A}{n}\right) + \cdots + \left(X_k - \frac{A}{n}\right) \geq cn - \epsilon n \,\right]$$

$$\leq \sum_i \Pr\left[\, X_i - \frac{A}{n} \geq (c - \epsilon)n \,\right]$$

$$\leq \sum_i \frac{1}{(c - \epsilon)^2 n^2} Var[X_i]$$

$$= \frac{1}{(c - \epsilon)^2 n^2} \sum_i Var[X_i]$$

$$\leq \frac{1}{(c - \epsilon)^2 n^2} kA$$

$$= \frac{\epsilon}{(c - \epsilon)^2}$$

Note that we have used Chebychev's inequality and union bounds in the above proof. We bound the error in the estimated value of A with the following lemma.

Lemma 4

Let $k \geq \epsilon n^2 / A$ and $d > 0$. Then

$$\Pr\left[\, |\frac{n}{k} \left(\sum_{i=1}^{k} X_i\right) - A| \geq dA \,\right] \leq \frac{1}{\epsilon d^2}$$

Proof

$$\Pr\left[\, |\frac{n}{k}\left(\sum_{i=1}^{k} X_i\right) - A| \geq t \, \right] = \Pr\left[\, |\left(\sum_{i=1}^{k} X_i\right) - \frac{k}{n}A| \geq \frac{kt}{n} \, \right]$$

$$= \Pr\left[\, |\left(\sum_{i=1}^{k} X_i - \frac{1}{n}A\right)| \geq \frac{kt}{n} \, \right]$$

$$\leq \frac{n^2}{k^2 t^2} k \cdot Var[X_i]$$

Let $k = \lambda \dfrac{n^2}{A}$, where $\lambda \geq \epsilon$. Then the above probability is less than or equal to

$$\frac{n^2}{k^2 t^2} k \cdot Var[X_i] \leq \frac{n^2}{\lambda \frac{n^2}{A} t^2} A$$

which is just $\dfrac{A^2}{\lambda t^2}$. Setting $Ad = t$ gives

$$\frac{1}{\lambda d^2} \leq \frac{1}{\epsilon d^2}$$

Random Dot Product Graph Models for Social Networks

Stephen J. Young[1] and Edward R. Scheinerman[2]

[1] Georgia Institute of Technology, Atlanta GA 30332-0160
[2] Johns Hopkins University, Baltimore MD 21218-2682

Abstract. Inspired by the recent interest in combining geometry with random graph models, we explore in this paper two generalizations of the random dot product graph model proposed by Kraetzl, Nickel and Scheinerman, and Tucker [1,2]. In particular we consider the properties of clustering, diameter and degree distribution with respect to these models. Additionally we explore the conductance of these models and show that in a geometric sense, the conductance is constant.

1 Introduction

With the ubiquity and importance of the Internet and genetic information in medicine and biology, the study of complex networks relating to the Internet and genetics continues to be an important and vital area of study. This is especially true for networks such as the physical layer of the Internet, the link structure of the world wide web, and protein-protein and protein-gene interaction networks. Due to the size of these networks [3] and the difficulty of determining complete link information [4,5] a significant amount of research has gone into finding models that match observed properties of these graphs in order to empirically (via simulation) and theoretically understand and predict properties of these complex networks. There are three models that, together with their variations, are the core models for these complex networks [6]. The configurational model and its variants attempt to generate complex networks by specifying the degree sequence and creating edges randomly with respect to that degree sequence. On the other hand, the Barabási-Albert preferential attachment model attempts to model the process by which the network grows, specifically, it posits that vertices with high degree are more likely to increase in degree when a new vertex is added to the network. In a similar vein, the copying model [7,8], also attempts to model the growth process of a complex networks. However, the copying model takes the more distinctly biological viewpoint of replication of existing nodes combined with mutation. All three of these types of models have had success in reproducing the hallmark features of complex networks, namely a power-law degree distribution, a diameter that grows slowly or is constant with the size of the graph, and one of several clustering properties; see [6,9] for a collection of such results.

However, there are many other aspects of complex networks that fail to be captured by these models, for example non-uniform assortativity [10] and the

A. Bonato and F.R.K. Chung (Eds.): WAW 2007, LNCS 4863, pp. 138–149, 2007.

existence of directed cycles, among others. Thus there is considerable interest in new models for complex networks that exhibit a power-law like degree sequence, small diameter, and clustering, and are different enough from the three main model classes to exhibit other properties of complex networks that are not exhibited by the current models. One potential method to create new models is to incorporate geometry into already existing models. Flaxman, et al. used geometry coupled with the preferential attachment model to create a model that generates a random power-law graph that has small separators [11].

Taking this idea one step further, one can add semantic information to an already existing model. One such model is the random dot product graph model applied by Caldarelli, et al. and Azar, et al. [12,13] and formalized by Kraetzl, Nickel, Scheinerman, and Tucker [1,2]. In their work they assign to each vertex a vector in $\mathbb{R}^d$ and then any edge is present with probability equal to the dot product of the endpoints. Thus, thinking of the vertices as members of a social network, the vectors together with the dot product encode semantically the idea of differing "interests" and varying levels of "talkativeness." We discuss the two natural generalizations of the random dot product graph model proposed by Kraetzl, et al, specifically, we remove the restrictions on the vectors imposed in their earlier work and develop directed generalization. First we briefly outline in Sect. 2 their model and the known results on diameter, clustering and degree distribution in order to provide a framework for the rest of this paper. We then present the two natural generalizations of the random dot product model. In Sect. 3 and Sect. 4 we demonstrate that an arbitrarily large fraction of the graph has constant diameter and that both the undirected and directed models demonstrate clustering. We derive in Sect. 5 explicit formulas for the degree sequence leading to a super-linear number of edges, which is consistent with recent results of Leskovec, Kleinberg and Faloutsos [14]. Finally, in Sect. 6 we turn our attention to conductance. We show that any small separators present are essentially non-semantic and leave open the question of general conductance. In Sect. 7 we discuss some areas for future work.

2   Model Specification

Kraetzl, Nickel and Scheinerman develop a new family of random graph model for social networks based on the dot product. In particular, they consider in detail the following model. Each vertex v is independently assigned a random vector, W_v, in $\mathbb{R}^d$, where each coordinate is independently and identically distributed as $\frac{1}{\sqrt{d}}\mathcal{U}^\alpha[0,1]$; that is a scaled copy the uniform distribution on $[0,1]$ to the α power. Then each edge $\{u,v\}$ is present independently with probability $\langle W_v, W_u \rangle$. They go on to show that the resulting graph G, has the following properties for $d = 1$:

1. The giant component of G has diameter almost surely at most 6 as $n \to \infty$.
2. For all vertices $u, v,$ and w,

$$\mathbb{P}\left(u \sim w \mid u \sim v \sim w\right) = \left(\frac{\alpha+1}{2\alpha+1}\right)^2 > \mathbb{P}\left(u \sim w\right) = \frac{1}{(\alpha+1)^2} \ . \quad (1)$$

3. The expected number of vertices of degree k on a n vertex graph generated in this manner is

$$\frac{1}{k!\alpha}(1+\alpha)^{\frac{1}{\alpha}}\Gamma\left(\frac{1}{\alpha}+k\right)n^{\frac{\alpha-1}{\alpha}} . \tag{2}$$

They proceed to show that for higher dimensions the probability of an arbitrary edge is independent of the dimension, but the degree distribution develops a "bend" in the power law. That is, the slope of the log-log plot of the degree distribution in numerical studies (and confirmed analytically for $d = 2$) decreases sharply for some given degree, which they conjecture to be $n/(d\alpha + d)$.

We consider the two natural generalizations of this model, one undirected and one directed, and show that they behave similarly to the model described by Kraetzl, Nickel and Scheinerman and resolve some of the higher dimensional questions posed regarding the nature of clustering and the diameter in the model.

First we consider the undirected generalization. Let $\mathbf{W}$ be a random variable on a $\mathbb{R}^d$ such that if W_i and W_j are distributed as $\mathbf{W}$, $\mathbb{P}\left(\langle W_i, W_j \rangle \in (0,1)\right) = 1$. Then we define $G(\mathbf{W}, n)$ as the graph on n vertices where each vertex v is assigned a vector W_v distributed as $\mathbf{W}$ and each edge $\{u, v\}$ is present independently with probability $\langle W_u, W_v \rangle$. It is clear from this construction that the restriction on the nature of the distribution $\mathbf{W}$ is necessary in order to guarantee that the inner products are all valid, nontrivial probabilities. When a distribution satisfies this condition, we shall say that it satisfies the inner product condition. Note that the inner product condition implies that $\mathbb{P}\left(\|\mathbf{W}\| < 1\right) = 1$, and guarantees that there is always some probability of an edge appearing (or not appearing) between any two pairs of vertices. Although it may seem more natural to allow for 0 or 1 inner products, precluding these values simplifies the analysis by forbidding pathological and uninteresting cases that can come about when there is a positive probability of guaranteeing or forbidding an edge.

The natural generalization of $G(\mathbf{W}, n)$ is to consider a directed graph with similar properties. Suppose $(\mathbf{X}, \mathbf{Y})$ is a pair of distributions on $\mathbb{R}^d \times \mathbb{R}^d$ such that if X_u is distributed as $\mathbf{X}$ and Y_v is distributed as $\mathbf{Y}$, $\mathbb{P}\left(\langle X_u, Y_v \rangle \in (0,1)\right) = 1$. We will abuse terminology slightly and say that such a $(\mathbf{X}, \mathbf{Y})$ pair satisfies the inner product condition. Then we consider the random directed graph $\overrightarrow{G}(\mathbf{X}, \mathbf{Y}, n)$ as the graph on n vertices, where each vertex v is assigned a pair of vectors (X_v, Y_v) and each directed edge (u, v) is present independently with probability $\langle X_u, Y_v \rangle$. Again, the inner product condition is a natural condition driven by the necessity for the quantity associated to an arc being a probability. Note that it is clearly not necessary for either of $\mathbf{X}$ or $\mathbf{Y}$ to have bounded norm, however we believe that the nature of those distributions such that $(\mathbf{X}, \mathbf{Y})$ satisfies the inner product condition and has unbounded norm are so pathological as to be uninteresting. Thus, for the remainder of this paper we assume that there is some compact set K such that $\mathbb{P}\left(\mathbf{X} \in K\right) = \mathbb{P}\left(\mathbf{Y} \in K\right) = 1$. Note as well that for clarity of presentation, we will abuse notation and say that a vertex belongs to a region R whenever its assigned vector(s) lie in that region.

We observe that $G(\mathbf{W}, n)$ generalizes both the Erdős-Rényi model and a version of the configurational random graph model. The first is achieved by letting

W be a constant random variable. Then it is clear that the model under consideration is just the Erdős-Rényi model with parameter $\langle \mathbf{W}, \mathbf{W} \rangle$. Also note that this holds for $\overrightarrow{G}(\mathbf{X}, \mathbf{Y}, n)$ by letting both **X** and **Y** be constant. Now by letting $d = 1$ and $\mathbb{P}(\mathbf{W} = k/c)$ be proportional to $k^{-\alpha}$, where c is a normalizing constant, we have a model that generalizes a randomized configurational model.

In addition to generalizing the Erdős-Rényi and configurational models, there is a natural interpretation of the vectors and the interaction of those vectors in the (directed) random dot product graph model. By considering each component of the vector associated with a vertex as a property or interest of that vertex, we may interpret the value of the component in a natural way. Furthermore, recent research into the nature of links in the blogosphere, specifically the Live Journal networks, have shown that a significant percentage of links can be explained by properties of the blog, such as the location of the author, interest lists, age, gender, etc. [15]. This interpretation of random dot product graphs provides a ready-made collection of tools for creating distributions by applying previous research into the singular value decomposition and related methods for feature extraction.

Just as representing entities as vectors, or pairs of vectors, is a natural idea, we feel that the inner product is a natural way of encapsulating two primary barriers to "linking". More explicitly, two websites are unlikely to have a direct link if their topics are completely unrelated, this corresponds to their vectors having a large angle between them in the dot product graph representation. On the other hand, if two websites have nearly identical topics, they still may not be linked due to the selectivity of one of the websites. That is, if one of the websites doesn't link to many things overall, then no matter how close another website's interests are there is still a significant barrier to "linking". The inner product encapsulates both these barriers in that both the angle between the vector and the norm of the vectors impact the inner (dot) product.

3 Diameter of "Giant" Component

In this section, we show that an arbitrarily large fraction of the graph generated by $G(\mathbf{W}, n)$ almost surely forms a connected graph with diameter at most 5. In a slight abuse of standard terminology, we will refer to this arbitrarily large fraction of the graph as the "giant" component. A key step in the proof of the diameter of the "giant" component for $\overrightarrow{G}(\mathbf{X}, \mathbf{Y}, n)$ is the following lemma, which generalizes the result on the diameter of the Erdős-Rényi random graph model.

Lemma 1. *Let D be a directed random graph on v vertices such that each directed edge is present independently with probability at least p. Then D is almost surely strongly connected with directed diameter 2.*

Proof. Consider some pair of vertices, u and v. The probability that there is not a directed path of length at most 2 from u to v is at most $(1-p^2)^{|V(D)|-2}(1-p)$. Thus the probability that u and v are not strongly connected by paths of length at most 2 is at most $1 - (1 - (1-p^2)^{n-2}(1-p))^2$. But then, the expected number of such pairs that are not strongly connected by paths of length at most 2 is at most

$$n(n-1)(2(1-p^2)^{n-2}(1-p) - (1-p^2)^{2n-4}(1-p)^2) \qquad (3)$$

which approaches 0 as $n \to \infty$. Thus D is almost surely strongly connected with directed diameter at most 2 [16].

We will denote by $B(c;r)$ (respectively $\overline{B}(c;r)$) the open (respectively closed) ball of radius r centered at c.

Theorem 1. *Let* $\mathbf{W}, \mathbf{X}, \mathbf{Y}$ *be distributions on* $\mathbb{R}^d$ *such that* $\mathbf{W}$ *and* $(\mathbf{X}, \mathbf{Y})$ *satisfy the inner product condition. Further assume that there is some compact region* K *such that* $\mathbf{X}$ *and* $\mathbf{Y}$ *lie inside* K *almost surely. Then an arbitrarily large fraction of* $G(\mathbf{W}, n)$ *is connected with diameter 5 and an arbitrarily large fraction of* $\overrightarrow{G}(\mathbf{X}, \mathbf{Y}, n)$ *is strongly connected with directed diameter at most 5.*

We prove only the undirected case here as the directed case follows a similar but more complicated argument.

Proof. We may assume without loss of generality that $\mathbf{W} \in \overline{B}(0;1)$. Letting $0 < \delta < \frac{1}{4}$, choose $\epsilon > 0$ such that $\mathbb{P}(\mathbf{W} \in B(0;\epsilon)) < \delta$. Then let A be the closed annulus $\overline{B}(0;1) - B(0;\epsilon)$. For all $\alpha \in A$, choose

$$r_\alpha \in \left\{ r > 0 \,\middle|\, \forall x, y \in B(\alpha;r), x^T y > \frac{\epsilon^2}{4} \right\}, \qquad (4)$$

which is non-empty by the continuity of the inner product. Then $\cup_{\alpha \in A} B(\alpha;r_\alpha)$ is an open cover of the compact set A with some finite subcover, say $\{B(\alpha_i;r_{\alpha_i})\}$.

Fix i such that $\mathbb{P}(\mathbf{W} \in B(\alpha_i;r_{\alpha_i})) \neq 0$. Then, as $n \to \infty$, there are almost surely infinitely many vertices that lie in $B(\alpha_i;r_{\alpha_i})$. It then follows from a result of Erdős and Renyi, since the probability of every edge is at least $\frac{\epsilon^2}{4}$ and for fixed $\{W_v\}$ each edge is present independently, the graph induced by $B(\alpha_i;r_{\alpha_i})$ has diameter at most 2, almost surely. Clearly, if $\mathbb{P}(\mathbf{W} \in B(\alpha_i;r_{\alpha_i})) = 0$, then there are almost surely no vertices in that region, and moreover those regions do not affect the diameter of $G(\mathbf{W}, n)$.

Now consider two regions $\mathcal{R}_i = B(\alpha_i;r_{\alpha_i})$ and $\mathcal{R}_j = B(\alpha_j;r_{\alpha_j})$ occurring with positive probability. There is a naturally defined probability measure on $\mathcal{R}_i \times \mathcal{R}_j$. Furthermore, since $\mathbb{P}(W_i^T W_j = 0) = 0$, there exist $\hat{\epsilon}, \hat{\delta} > 0$ such that $\mathbb{P}(W_i^T W_j > \hat{\delta} \mid W_i \in \mathcal{R}_i, W_j \in \mathcal{R}_j) > \hat{\epsilon}$. But, since $\hat{\delta}$ and $\hat{\epsilon}$ are independent of n, and since $\mathcal{R}_i$ and $\mathcal{R}_j$ almost surely contain an infinite number of vertices; there is almost surely an edge between the regions. Now given vertices $u \in \mathcal{R}_i$ and $v \in \mathcal{R}_j$, there is almost surely an edge e between $\mathcal{R}_i$ and $\mathcal{R}_j$, a path of length 2 from u to e, and a path of length 2 from e to v. Thus, for any pair of vertices in A there is almost surely a path of length at most 5 between them. But A asymptotically contains $(1 - \delta)n$ vertices, and since δ was arbitrary, A contains an arbitrarily large fraction of the vertices.

4   Clustering

In this section, we examine the clustering of $G(\mathbf{W}, n)$ and $\overrightarrow{G}(\mathbf{X}, \mathbf{Y}, n)$ and find that except in the case of constant random variables, the presence clustering is independent of the random variables. In order to show the clustering results we need the following convexity result, which will allow the use of Jensen's Inequality in the proof of Theorem 2.

Lemma 2. *Let $a, b \in \mathbb{R}^d$. Let $D \subseteq \mathbb{R}^d$ be a region such that for all $x \in D$, $\langle a, x \rangle \in (0, 1)$ and $\langle b, x \rangle \in (0, 1)$. Then $u \colon D \longrightarrow \mathbb{R}$ defined by $x \longmapsto \langle a, x \rangle \langle b, x \rangle$ is a convex function of x.*

Proof. Let $F : (0, 1) \times (0, 1) \longrightarrow \mathbb{R}$ be defined by $(x, y) \longmapsto xy$. We note that $\nabla^2 F = \begin{pmatrix} 0 & 1 \\ 1 & 0 \end{pmatrix}$. This matrix, although not positive semi-definite, is positive semidefinite over $[0, 1] \times [0, 1]$, and hence $F(x, y)$ is convex over its domain [17]. Now note that since $\langle a, x \rangle$ is a real inner product, for any $\lambda \in [0, 1]$ and $x, y \in D$, $\langle a, \lambda x + (1 - \lambda)y \rangle = \lambda \langle a, x \rangle + (1 - \lambda) \langle a, y \rangle$. Thus $\langle a, x \rangle$ is a convex function in x and similarly for $\langle b, x \rangle$. Thus $u(x) = F(\langle a, x \rangle, \langle b, x \rangle)$ is the composition of convex functions and hence is convex.

Theorem 2. *Let $W_v, W_w, W_u, X_u, X_v, X_w, Y_u, Y_v, Y_w$ be independent random variables distributed over $\mathbb{R}^d$, not necessarily identically distributed, such that $\langle W_i, W_j \rangle$ and $\langle X_i, Y_j \rangle$ satisfy the inner product condition for all $i \neq j$. For the undirected graph where each edge $\{i, j\}$ is present with probability $\langle W_i, W_j \rangle$, we have that*

$$\mathbb{P}\left(u \sim v \mid u \sim v, v \sim w\right) \geq \mathbb{P}\left(u \sim v\right). \tag{5}$$

Now consider the random directed graph where each arc $i \to j$ is present, independently, with probability $\langle X_i, Y_j \rangle$. Then we have that

1. $\mathbb{P}\left(u \to w \mid u \to v, v \to w\right) \geq \mathbb{P}\left(u \to w\right)$,
2. $\mathbb{P}\left(u \to w \mid u \to v, w \to v\right) \geq \mathbb{P}\left(u \to w\right)$,
3. $\mathbb{P}\left(u \to w \mid v \to u, v \to w\right) \geq \mathbb{P}\left(u \to w\right)$, *and*
4. $\mathbb{P}\left(u \to w \mid w \to v, v \to u\right) = \mathbb{P}\left(u \to w\right)$.

As an immediate corollary, we get that for any set of vertices u, v and w in $G(\mathbf{W}, n)$, we get $\mathbb{P}\left(u \sim v \mid u \sim v, v \sim w\right) \geq \mathbb{P}\left(u \sim v\right)$ and for any set of vertices u, v and w in $\overrightarrow{G}(\mathbf{X}, \mathbf{Y}, n)$

1. $\mathbb{P}\left(u \to w \mid u \to v, v \to w\right) \geq \mathbb{P}\left(u \to w\right)$,
2. $\mathbb{P}\left(u \to w \mid u \to v, w \to v\right) \geq \mathbb{P}\left(u \to w\right)$,
3. $\mathbb{P}\left(u \to w \mid v \to u, v \to w\right) \geq \mathbb{P}\left(u \to w\right)$, and
4. $\mathbb{P}\left(u \to w \mid w \to v, v \to u\right) = \mathbb{P}\left(u \to w\right)$.

Note that in $G(\mathbf{W}, n)$, equality holds bounds on clustering if and only if $\mathbf{W}$ is a constant random variable.

5   Degree Distribution

We derive, in this section, a set of natural formulas for the degree distributions of both $G(\mathbf{W}, n)$ and $\overrightarrow{G}(\mathbf{X}, \mathbf{Y}, n)$. In Sect. 5.1, we discuss the application of these formulas to the construction of specific random models meeting desired degree sequence considerations.

Proposition 1. *Let $G = G(\mathbf{W}, n)$ where $\mathbf{W}$ satisfies the inner product condition and let $D = \overrightarrow{G}(\mathbf{X}, \mathbf{Y}, n)$ where $\mathbf{X}$ and $\mathbf{Y}$ are distributions over $\mathbb{R}^d$ where $(\mathbf{X}, \mathbf{Y})$ satisfies the inner product condition. Then, for a vertex $w \in V(G)$*

$$\mathbb{P}\left(\deg(w) = k\right) = \int \binom{n-1}{k} \langle \mathbb{E}\left[\mathbf{W}\right], W \rangle^k \left(1 - \langle \mathbb{E}\left[\mathbf{W}\right], W \rangle\right)^{n-k-1} d\mathbf{W}. \quad (6)$$

Furthermore, for a vertex $v \in V(D)$

$$\mathbb{P}\left(\deg^+(v) = k\right) = \int \binom{n-1}{k} \langle \mathbb{E}\left[\mathbf{X}\right], Y \rangle^k \left(1 - \langle \mathbb{E}\left[\mathbf{X}\right], Y \rangle\right)^{n-1-k} d\mathbf{Y} \quad (7)$$

$$\mathbb{P}\left(\deg^-(v) = k\right) = \int \binom{n-1}{k} \langle \mathbb{E}\left[\mathbf{Y}\right], X \rangle^k \left(1 - \langle \mathbb{E}\left[\mathbf{Y}\right], X \rangle\right)^{n-1-k} d\mathbf{X}. \quad (8)$$

This leads to an immediate result on the density of edges in $G(\mathbf{W}, n)$ and $\overrightarrow{G}(\mathbf{X}, \mathbf{Y}, n)$.

Corollary 1. *Let $G = G(\mathbf{W}, n)$ where $\mathbf{W}$ satisfies inner product condition and let $D = \overrightarrow{G}(\mathbf{X}, \mathbf{Y}, n)$ where $\mathbf{X}$ and $\mathbf{Y}$ are distributions over $\mathbb{R}^d$ where $(\mathbf{X}, \mathbf{Y})$ satisfies the inner product condition. Then $\mathbb{E}\left[|E(G)|\right] = \binom{n}{2} \langle \mathbb{E}\left[\mathbf{W}\right], \mathbb{E}\left[\mathbf{W}\right] \rangle$ and $\mathbb{E}\left[|E(D)|\right] = n(n-1) \langle \mathbb{E}\left[\mathbf{X}\right], \mathbb{E}\left[\mathbf{Y}\right] \rangle$.*

This implies that the edge density is $\Omega(n^2)$, contrary to conventional wisdom regarding complex networks. However we feel that this trade off is acceptable in practice for several reasons. The first being that $\langle \mathbb{E}\left[\mathbf{W}\right], \mathbb{E}\left[\mathbf{W}\right] \rangle$ and $\langle \mathbb{E}\left[\mathbf{X}\right], \mathbb{E}\left[\mathbf{Y}\right] \rangle$ are typically small. Furthermore, although the results regarding the diameter of the graph would not hold, one could consider $\mathbf{X}$ and $\mathbf{Y}$ as functions of n and introduce sparsity in that manner. We also note that, particularly for the world wide web, gene-protein networks, and the Internet, it is widely accepted that empirical studies are not capturing all the edges present. Combine this fact with recent work showing that the incompleteness can severely skew some statistics of the data [4,5], and it is plausible that one or more of these networks is not truly sparse. In addition, the recent work of Leskovec, Kleinberg and Faloutsos [14] has shown that for many social networks the number of edges is becoming super-linear in the number of vertices as these networks evolve.

5.1   Sample Distributions

Although it is obvious that not every distribution $\mathbf{W}$ or pair of distributions $(\mathbf{X}, \mathbf{Y})$ can lead to a power law, it is useful to discuss a few means of generating power law degree distributions. We will focus on the directed model, as Kraetzl, Nickel and Scheinerman have already shown one manner in which to achieve a power law degree distribution for the undirected model [1].

It is natural to consider directed versions of complex network where the in-degrees are distributed as a power law, while the out-degrees tend to be more concentrated, in order to capture situations where the for physical reasons the the out-degree is limited. Thus since we know that the Erdős-Rényi graph model tends to produce a concentrated degree sequence and further if each component is independently distributed as $\frac{1}{\sqrt{d}}\mathcal{U}^\alpha(0,1)$ in the undirected random dot product graph model tends to produce a power law, it is natural to attempt to emulate these two in the directed model. Thus, taking each component of $\mathbf{X}$ to be independently distributed as $\frac{1}{\sqrt{d}}\mathcal{U}(0,1)$ (that is, having low variance, similarly to the Erős-Rényi model) and each component of $\mathbf{X}$ to be $\frac{1}{\sqrt{d}}\mathcal{U}^\alpha(0,1)$ (and thus having high variance), with $\alpha = 16$, $d = 5$, $n = 10000$, and with 200 trials, yields the average degree distributions shown in Fig. 5.1. Note that this roughly models the desired behavior, in that the out-degree is strongly clustered around a single value and the in-degrees are distributed as a power-law. Further empirical refinement can lead to a closer approximation, and thus through repeated simulation and tuning it is reasonable to assume that this degree distribution and others like it, can be well approximated.

We now note that for any orthonormal matrix Q and any non-zero constant c, $\vec{G}(\mathbf{X}, \mathbf{Y}, n) = \vec{G}(cQ\mathbf{X}, \frac{1}{c}Q\mathbf{Y}, n)$. Thus we may assume that

1. $\langle \mathbb{E}[\mathbf{X}], e_1 \rangle = \langle \mathbb{E}[\mathbf{Y}], e_1 \rangle = \sqrt{\langle \mathbb{E}[\mathbf{X}], \mathbb{E}[\mathbf{Y}] \rangle}$,
2. $\langle \mathbb{E}[\mathbf{X}], e_2 \rangle \geq 0$,
3. $\langle \mathbb{E}[\mathbf{Y}], e_2 \rangle = 0$, and
4. $\langle \mathbb{E}[\mathbf{X}], e_i \rangle = \langle \mathbb{E}[\mathbf{Y}], e_i \rangle = 0$ for $i > 2$.

In particular, combining these observations with (7) and (8), we obtain moments for some components of $\mathbf{X}$ and $\mathbf{Y}$ if $\vec{G}(\mathbf{X}, \mathbf{Y}, n)$ satisfies a given degree distribution. However, these moments do not fully characterize $\mathbf{X}$ and $\mathbf{Y}$, but rather limit the space of feasible distributions.

Perhaps more useful from a modeling point of view is the possibility of using Kernel Density Estimators of Hörmann and Leydold [18] to develop estimated distributions for $(\mathbf{X}, \mathbf{Y})$. In particular, given a graph G and a vector $\{V_i\}$ for each vertex so that G is "generated" by $\{V_i\}$ under the random dot product graph model, Kernel Density Estimation provides a means to rapidly generate approximate samples from the sample distribution. (For more details on extracting vectors from a given graph see [19,20,21].) Thus using $\{V_i\}$ it is theoretically possible to generate a random graph that "looks-like" G.

6 On the Nature of Bad Cuts

In light of the work of Flaxman, Frieze and Vera which showed that the geometric preferential attachment graph has bad cuts that are due to the geometry

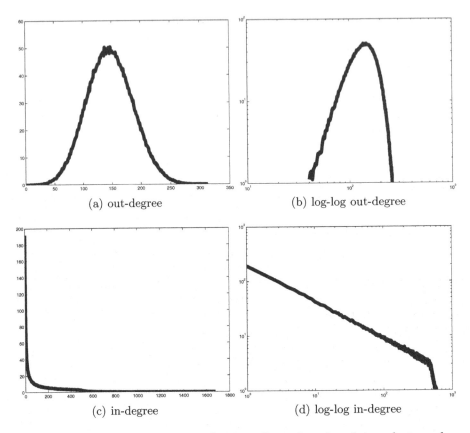

(a) out-degree

(b) log-log out-degree

(c) in-degree

(d) log-log in-degree

Fig. 1. Average degree sequences for given directed random dot product graph

of the underling space [11], it is natural to consider whether the random dot product graphs exhibit similar behavior. In this section we characterize the conductance of the geometric cuts in the undirected random dot product graph model. Specifically, we show that any bad cuts have no semantic content. That is, if low conductance cuts exist they are essentially non-geometric.

Before discussing the nature of the geometric cuts in $G(\mathbf{W}, n)$ we first need some preliminary definitions. For any region R we will abuse notation and refer to the set of vertices whose vectors are contained in the region as R. We will also, for notational convenience, denote by W_R the expectation of $\mathbf{W}$ restricted to R for any $\mathbf{W}$-measurable set R. Using notation standard from conductance we will denote by $\mathrm{Vol}(R)$ the sum of the degrees for all vertices in R and by $C(R, \overline{R})$ the number of edges crossing the cut $(R, \overline{R})$. Finally, we also denote by $\mathbb{P}(R)$ the probability that a random variable distributed as $\mathbf{W}$ lies within the region R. The conductance of the cut $(R, \overline{R})$, $\Phi_R(G(\mathbf{W}, n))$, is defined as $C(R, \overline{R}) / \min \left\{ \mathrm{Vol}\, R, \mathrm{Vol}\, \overline{R} \right\}$. With these definitions and a multidimensional generalization of the Chernoff Bound [22], we have the following results on the nature of geometric cuts in $G(\mathbf{W}, n)$.

Theorem 3. *Let R be a fixed subset of $\mathbb{R}^d$ and let $\mathbf{W}$ be a distribution on $\mathbb{R}^d$ that satisfies the inner product condition. Then almost surely*

$$\lim_{n \to \infty} \Phi_R(G(\mathbf{W}, n)) \geq \frac{\mathbb{P}\left(\overline{R}\right) \langle W_R, W_{\overline{R}} \rangle}{\langle W_R, \mathbb{E}\left[\mathbf{W}\right] \rangle}, \tag{9}$$

when $\mathbb{P}\left(R\right) \|W_R\|^2 \leq \mathbb{P}\left(\overline{R}\right) \|W_{\overline{R}}\|^2$.

This results establishes that any fixed region does not induce a bad cut. However it leaves open the possibility that there is some sequence of regions giving arbitrarily small conductance. That is, it may be possible that for an arbitrary $c > 0$ there is some region R_c such that the conductance induced by the cut $(R_c, \overline{R}_c)$ is constant but less than c. In fact, by using the inner product condition we may show the following result:

Theorem 4. *For a fixed distribution $\mathbf{W}$ satisfying the inner product condition, $\inf_R \lim_{n \to \infty} \Phi_R(G(\mathbf{W}, n))$ is bounded below, where the infimum is taken over $\mathbf{W}$-measurable sets R.*

By combining the results of Theorem 3 and Theorem 4, we conclude that if $\mathbf{W}$ satisfies the inner product condition there is some $\alpha > 0$, depending only on $\mathbf{W}$, such that for any region $R \subseteq \mathbb{R}^d$, $\lim_{n \to \infty} \Phi_R(G(\mathbf{W}, n)) > \alpha$, with high probability. Specifically, any fixed partition $(R, \overline{R})$ of $\mathbb{R}^d$ has constant conductance independent of $(R, \overline{R})$. Thus, in contrast to the work of Flaxman, Frieze and Vera, where they showed that the geometric preferential attachment model has bad cuts induced entirely by the geometry, if the random dot product graph model has bad cuts they are entirely non-geometric. This does, however, leave open the question of what happens for non-fixed geometric regions and non-geometric partitions. Although we believe that the conductance of the random dot product graph model is asymptotically constant, the slow rate of convergence of this result leaves open the possibility that for every n there is a positive probability that some region has conductance smaller than α. Furthermore, since this result is inherently geometric, it says little about the case where $\mathbf{W}$ is not a continuous distribution. For instance, if $\mathbf{W}$ contains a point mass, then there is no way to geometrically place vertices generated by a point mass on opposing sides of a partition, whereas a partition of the vertices can clearly separate those vertices. Thus, fully resolving the conductance of the random dot product graph model will require a fundamental non-geometric insight into the structure of these graphs as in the work of Mihail, et al. [23].

7 Future Work

There are some natural questions that this work brings up. Perhaps the most pressing is the development of a sparse, or preferably, a variable density analogue of both $G(\mathbf{W}, n)$ and $\overrightarrow{G}(\mathbf{X}, \mathbf{Y}, n)$. Although, as we noted above, the presence of $\Omega\left(n^2\right)$ edges is not as major an objection as it once was for social networks, it still limits the models' general applicability. Thus a natural sparse generalization would broaden the applicability of these models. We have positive results in a preliminary work in this direction [24].

Also, given that the result that reinvigorated the study of social networks was Milgram's experimental result on the navigability of the "real world" social network, [25,26], it is reasonable to consider under what conditions short paths can be found in $G(\mathbf{W}, n)$. Kleinberg's result on the navigation of the grid with power-law shortcuts showed that navigation is sensitive to parameters of the model [27], however we feel that the additional semantic information in $G(\mathbf{W}, n)$ will allow navigation under more general conditions.

Finally, from a simulation point of view, it would be desirable to have a means of rapidly generating samples from $G(\mathbf{W}, n)$ or $\overrightarrow{G}(\mathbf{X}, \mathbf{Y}, n)$. Kraetzl, Nickel, and Scheinerman [1] discuss a thresholding modification of the natural means of generation that will produce an approximate sample. However, it is not immediately obvious how much the loss of edges due to thresholding will affect any given property of the sample graph. Thus, for serious simulation purposes, some means of estimating the effect of the thresholding or a clever way of reducing the overall computation time would seem to be necessary.

Acknowledgments. The first author is grateful to Milena Mihail for her advice and support in the preparation of this document. The authors would also like to thank the anonymous referees for the helpful comments.

References

1. Kraetzl, M., Nickel, C., Scheinerman, E.R.: Random dot product graphs: A model for social netowrks. Preliminary Manuscript (2005)
2. Kraetzl, M., Nickel, C., Scheinerman, E.R., Tucker, K.: Random dot product graphs (July 2005), http://www.ipam.ucla.edu/abstract.aspx?tid=5498
3. Albert, R., Barabási, A.L.: Statistical mechanics of complex networks. Rev. Modern Phys. 74(1), 47–97 (2002)
4. Achlioptas, D., Kempe, D., Clasuet, A., Moore, C.: On the bias of traceroute sampling or, power-law degree distributions in regular graphs. In: STOC 2005. Proc. of the 37th ACM Symposium on the Theory of Computer Science (2005)
5. Lakhina, A., Byers, J.W., Crovella, M., Xie, P.: Sampling biases in IP topology measurements. In: INFOCOM 2003. 22nd Joint Conference of the IEEE Computer and Communications Societies (2003)
6. Durrett, R.: Random graph dynamics. In: Cambridge Series in Statistical and Probabilistic Mathematics, Cambridge University Press, Cambridge (2007)
7. Chung, F., Galas, D.J., Dewey, T.G., Lu, L.: Duplication models for biological networks. Journal of Computational Biology (2003)
8. Kumar, R., Raghavan, P., Rajagopalan, S., Sivakumar, D., Tompkins, A., Upfal, E.: The web as a graph. In: PODS 2000. Proc. of the 19th ACM SIGMOD-SIGACT-SIGART Symposium on Principles of Database Systems, pp. 1–10. ACM Press, New York (2000)
9. Bornholdt, S., Schuster, H.G. (eds.): Handbook of graphs and networks. From the genome to the internet. Wiley-VCH, Weinheim (2003)

10. Newman, M.E.J.: Assortative mixing in networks. Physical Review Letters 89 (2002)
11. Flaxman, A.D., Frieze, A.M., Vera, J.: A geometric preferential attachment model of networks. Internet Math. 3(2), 187–205 (2006)
12. Caldarelli, G., Capocci, A., de Los Rios, P., Muñoz, M.A.: Scale-Free Networks from Varying Vertex Intrinsic Fitness. Physical Review Letters 89(25) (2002)
13. Azar, Y., Fiat, A., Karlin, A., McSherry, F., Saia, J.: Spectral analysis of data. In: STOC 2001. Proc. of the 33rd ACM Symposium on Theory of Computing, pp. 619–626. ACM Press, New York (2001)
14. Leskovec, J., Kleinberg, J., Faloutsos, C.: Graph evolution: Densification and shrinking diameters. ACM Trans. Knowl. Discov. Data 1(1) (2007)
15. Liben-Nowell, D., Novak, J., Kumar, R., Raghavan, P., Tomkins, A.: Geographic routing in social networks. Proceedings of the National Academy of Sciences 102(33), 11623–1162 (2005)
16. Bollobás, B.: Modern graph theory. In: Bollobás, B. (ed.) Graduate Texts in Mathematics, vol. 184, Springer, New York (1998)
17. Ben-Tal, A., Nemirovski, A.: Lectures on Modern Convex Optimization; Analysis, Algorithms, and Engineering Applications, SIAM, Philadelphia, PA (2001)
18. Hörmnn, W., Leydold, J.: Random-number and random-variate generation: automatic random variate generation for simulation input. In: Winter Simulation Conference, pp. 675–682 (2000)
19. Scheinerman, E.R., Tucker, K.: Exact and asymptotic dot product representations of graphs i: Fundamentals (Submitted, 2007)
20. Scheinerman, E.R., Tucker, K.: Exact and asymptotic dot product representations of graphs ii: Characterization and recognition (Submitted, 2007)
21. Scheinerman, E.R., Tucker, K.: Modelling graphs using dot product representations. (preparation, 2007)
22. Alon, N., Spencer, J.H.: The Probabilistic Method. In: Wiley-Interscience Series in Discrete Mathematics and Optimization, 2nd edn., Wiley-Interscience, New York (2000)
23. Mihail, M., Papadimitriou, C., Saberi, A.: On certain connectivity properties of the internet topology. J. Comput. System Sci. 72(2), 239–251 (2006) (FOCS 2003 Special Issue)
24. Young, S.J.: Sparse random dot product graphs. (preparation, 2007)
25. Milgram, S.: The small world problem. Psychology Today (1967)
26. Milgram, S., Travers, J.: An experimental study of the small world problem. Sociometry 32(4), 425–443 (1969)
27. Kleinberg, J.M.: The small world phenomenon: an algorithmic perspective. In: STOC 1999. Proc. of the 32nd ACM Symposium on the Theory of Computer Science (1999)

Local Computation of PageRank Contributions

Reid Andersen[1], Christian Borgs[2], Jennifer Chayes[2], John Hopcraft[3],
Vahab S. Mirrokni[2], and Shang-Hua Teng[4]

[1] University of California at San Diego, San Diego, CA
randerse@math.ucsd.edu
[2] Microsoft Research, Redmond, WA
{borgs,jchayes,mirrokni}@microsoft.com
[3] Cornell University, Ithaca, NY
jeh@cs.cornell.edu
[4] Boston University, Boston, MA
steng@cs.bu.edu

Abstract. Motivated by the problem of detecting link-spam, we consider the following graph-theoretic primitive: Given a webgraph G, a vertex v in G, and a parameter $\delta \in (0,1)$, compute the set of all vertices that contribute to v at least a δ fraction of v's PageRank. We call this set the δ-contributing set of v. To this end, we define the contribution vector of v to be the vector whose entries measure the contributions of every vertex to the PageRank of v. A local algorithm is one that produces a solution by adaptively examining only a small portion of the input graph near a specified vertex. We give an efficient local algorithm that computes an ϵ-approximation of the contribution vector for a given vertex by adaptively examining $O(1/\epsilon)$ vertices. Using this algorithm, we give a local approximation algorithm for the primitive defined above. Specifically, we give an algorithm that returns a set containing the δ-contributing set of v and at most $O(1/\delta)$ vertices from the $\delta/2$-contributing set of v, and which does so by examining at most $O(1/\delta)$ vertices. We also give a local algorithm for solving the following problem: If there exist k vertices that contribute a ρ-fraction to the PageRank of v, find a set of k vertices that contribute at least a $(\rho - \epsilon)$-fraction to the PageRank of v. In this case, we prove that our algorithm examines at most $O(k/\epsilon)$ vertices.

1 Introduction

In numerous applications of PageRank one needs to know, in addition to the rank of a given web page, which pages or sets of pages contribute most to its rank. These PageRank contributions have been used for link spam detection [4,10] and in the classification of web pages [12]. A set of pages that contributes significantly to the PageRank of a page is often called a *contribution set* or *supporting set* of the page [4,10].

The contribution that a vertex u makes to the PageRank of a vertex v is defined rigorously in terms of personalized PageRank. For a webgraph $G = (V, E)$ and a *teleportation constant* α (sometimes called the restart probability),

A. Bonato and F.R.K. Chung (Eds.): WAW 2007, LNCS 4863, pp. 150–165, 2007.
© Springer-Verlag Berlin Heidelberg 2007

let $\mathbf{PRM}_\alpha$ be the matrix whose u^{th} row is the personalized PageRank vector of u. The PageRank contribution of u to v, written $\mathrm{pr}_\alpha(u \to v)$, is defined to be the entry (u, v) of this matrix. The PageRank of a vertex v is the sum of the vth column of the matrix $\mathbf{PRM}_\alpha$, and thus the PageRank of a vertex can be viewed as the sum of the contributions from all other vertices. The *contribution vector* of v is defined to be the vth column of the matrix $\mathbf{PRM}_\alpha$, whose entries are the contributions of every vertex to the PageRank of v.

Given that the web graph is massive and getting larger at a substantial rate, it is essential to compute contribution vectors and identify supporting sets by examining as small a fraction of the graph as possible. In particular, it is helpful to design a *local* algorithm for computing the supporting sets of a particular vertex. Local algorithms search for a solution near a specified vertex by adaptively examining only a small subset of the input graph. They have been studied previously in distributed computing [16] and in graph partitioning and clustering [20,2]. Personalized PageRank vectors can be approximated locally. Using one of several possible algorithms [14,5,19], it is possible to compute an approximation of the personalized PageRank vector of a vertex u by examining only $O(1/\epsilon)$ vertices, where ϵ is the desired amount of error at each vertex.

Problem Formulation. Inspired by local algorithms for computing personalized PageRank, and motivated by the importance of supporting sets in link-spam detection, we consider the problem of directly computing the contribution vector of a given vertex to quickly identify its supporting sets. In particular, we consider following graph-theoretic primitive: Given a webgraph G, a vertex v in G, and a parameter $\delta \in (0,1)$, compute the set of all vertices each contributing at least a δ fraction to the PageRank of v. We call this set the δ-*contributing set of v*.

Such a primitive is useful for spam detection, since, given a webpage whose PageRank has recently increased suspiciously, we can quickly identify the set of pages that contribute significantly to the PageRank of that suspicious page. The above primitive may also be useful for analyzing social networks. In social networks in which the links capture the influence of vertices on each other, we can identify the nodes with the most influence to a given node.

Our Results. We give an efficient, local algorithm for computing an ϵ-approximation of the contribution vector for a given vertex v, a vector whose difference from the contribution vector is at most ϵ at each vertex. We prove that the number of the vertices examined by the algorithm is $O(1/\epsilon)$. The algorithm performs a sequence of probability-pushing operations on vertices of the graph, which we call pushback operations. When the pushback operation is applied to a vertex u, we perform a small amount of computation for each in-neighbor of u. Particularly, we add a fraction of a number stored at u to a number stored at each in-neighbor of u. The number of such operations that our algorithm performs is $O(1/\epsilon)$, and its running time can be bounded by the sum of the in-degrees of the vertices from which these operations were performed. To derive this algorithm, we adapt Jeh and Widom's technique for computing personalized PageRank vectors [14] to directly compute contribution vectors. To analyze the algorithm's running time and error bounds, we use techniques developed for the local clustering algorithm in [2].

Using our algorithm for approximating contribution vectors, we give an approximation algorithm to the primitive defined above. Explicitly, we give a local algorithm that returns a set containing the δ-contributing set of v and at most $O(1/\delta)$ vertices from the $\delta/2$-contributing set of v. Our algorithm applies at most $O(1/\delta)$ pushback operations. We also give a local algorithm for solving the following problem: If there are k vertices which contribute a ρ-fraction to the PageRank of v, find a set of k vertices which contribute at least $(\rho - \epsilon)$-fraction to the PageRank of v. In this case, we prove that our algorithm needs at most $O(k/\epsilon)$ pushback operations.

Finally, we remark that, in principle, one could directly compute the contribution vector for a vertex v by approximating the personalized PageRank vector of v in the time-reversal of the random walk Markov chain. We describe the computation required for this approach, and argue that for most graphs it is not as efficient as the method we propose.

Related Work. Supporting sets and PageRank contributions have been studied before as a tool for spam detection, notably in the SpamRank algorithm of Benczúr et al. [4], and in the Spam Mass algorithm of Gyöngyi et al. [10]. However, none of these papers developed a local algorithm for computing the contribution vector or supporting set. In the SpamRank algorithm [4], the contribution vectors are computed in the following way. One computes an approximation of each personalized PageRank vector in the graph to create an approximate PageRank matrix, and then takes the transpose of this matrix to obtain the approximate contribution vectors. This method is efficient for the task of computing the contribution vectors for every vertex in the graph, and it leverages fast algorithms for computing many personalized PageRank vectors simultaneously [9,19], but it does not provide an efficient way to compute the contribution vectors of a few selected suspicious vertices. Furthermore, the relative error in the resulting approximate contribution vectors may be larger than the relative error in the computed personalized PageRank vectors, since this is not preserved by the transpose operation.

PageRank contributions have also been used to estimate the PageRank of a target vertex. The algorithm in [7] heuristically identifies the top contributors to a vertex v by adaptively choosing vertices with high likelihood of being large contributors, and then locally computes personalized PageRank from those vertices. This is different from our approach of directly computing the contribution vector, and more difficult to analyze rigorously.

Local algorithms have been studied in distributed computing [16] and in graph partitioning and clustering [20,2]. Personalized PageRank vectors can be computed locally using a number of methods [5,2,19], many of which are based on the algorithm of Jeh and Widom [14]. None of these algorithms can be used directly to compute a contribution vector or supporting set.

There are numerous methods for detecting link spam besides the SpamRank-type algorithms we have mentioned here. Examples include applying machine learning to link-based features [3], the analysis of page content [15,17], TrustRank [11] and Anti-TrustRank [18], and statistical analysis of various page features [8]. Finally, in a follow-up to this paper we use the local algorithm developed here to design several locally computable page features for link spam detection, and evaluate these features experimentally [1].

Organization. This paper will be organized as follows. In Section 2, we review the basic concepts used in this paper, including PageRank, personalized PageRank, and PageRank contribution vectors. In Section 3, we derive an alternate formula for the PageRank contribution vector. Using this formula, we present an efficient local algorithm for computing PageRank contribution and analyze its performance. In Section 4, we consider several notions of supporting sets, which are sets of vertices that contribute significantly to the PageRank of a target vertex, and show how to efficiently compute approximate supporting sets. In Section 5 we make a few concluding remarks. We also show that, in principle, the time-reverse Markov chain can be used to compute the contribution vector, but argue that our method is more efficient.

2   Preliminaries

The web can be modeled by a directed graph $G = (V, E)$ where V are webpages and a directed edge $(u \rightarrow v) \in E$ represents a hyperlink in u that references v. Although the web graph is usually viewed as an unweighted graph, our discussion can be extended to weighted models. To deal with the problem of dangling nodes with no out-edges, we assume an artificial node with a single self-loop has been added to the graph, and an edge has been added from each dangling node to this artificial node. Let A denote the adjacency matrix of G. For each $u \in V$, let $d_{out}(u)$ denote the out-degree of u and let $d_{in}(u)$ denote the in-degree of u. Let D_{out} be the diagonal matrix of out-degrees.

We will now define PageRank vectors and contribution vectors. For convenience, we will view all vectors as row vectors, unless explicitly stated otherwise.

For a teleportation constant α, the PageRank vector $\mathbf{pr}_\alpha$ defined by Brin and Page [6] satisfies the following equation:

$$\mathbf{pr}_\alpha = \alpha \cdot \mathbf{1} + (1 - \alpha) \cdot \mathbf{pr}_\alpha \cdot M, \qquad (1)$$

where M is the random walk transition matrix given by $M = D_{out}^{-1} A$ and $\mathbf{1}$ is the row vector of all 1's (always of proper size). The PageRank of a page u is then $\mathrm{pr}_\alpha(u)$. When there is no danger of confusion, we may drop the subscript α. Note that the above definition corresponds to the normalization $\sum_u \mathrm{pr}_\alpha(u) = |V|$.

Similarly, the personalized PageRank vector $\mathbf{ppr}(\alpha, u)$ of a page $u \in V$, defined by Haveliwala [13], satisfies the following equation.

$$\mathbf{ppr}(\alpha, u) = \alpha \cdot \mathbf{e}_u + (1 - \alpha) \cdot \mathbf{ppr}(\alpha, u) \cdot M, \qquad (2)$$

where $\mathbf{e}_u$ is the row unit vector whose u^{th} entry is equal to 1.

Let $\mathbf{PRM}_\alpha$ denote the (personalized) PageRank matrix, whose uth row is the personalized PageRank vector $\mathbf{ppr}(\alpha, u)$. The (global) PageRank vector $\mathbf{pr}_\alpha$ is then $1 \cdot \mathbf{PRM}_\alpha$, the sum of all the personalized PageRank vectors. The *PageRank contribution* of u to v is defined to be the (u, v)th entry of $\mathbf{PRM}_\alpha$, and will be written $\mathrm{ppr}_\alpha(u \to v)$. The contribution vector $\mathbf{cpr}(\alpha, v)$ for the vertex v is defined to be the row vector whose transpose is the vth column of $\mathbf{PRM}_\alpha$. If $\mathbf{c} = \mathbf{cpr}(\alpha, v)$ is the contribution vector for v, then we denote by $\mathbf{c}(S)$ the total contribution of the vertices in S to the PageRank of v. In particular, we have $\mathbf{c}(V) = \mathrm{pr}_\alpha(v)$ and $\mathbf{c}(u) = \mathrm{ppr}_\alpha(u \to v)$.

3   Local Approximation of PageRank Contributions

In this section, we describe an algorithm for computing an approximation of the contribution vector $\mathbf{c} = \mathbf{cpr}(\alpha, v)$ of a vertex v.

Definition 1 (Approximate Contribution). *A vector $\tilde{\mathbf{c}}$ is an ϵ-approximation of the contribution vector $\mathbf{c} = \mathbf{cpr}(\alpha, v)$ if $\tilde{\mathbf{c}} \geq 0$ and, for all vertices u,*

$$\mathbf{c}(u) - \epsilon \cdot \mathrm{pr}_\alpha(v) \leq \tilde{\mathbf{c}}(u) \leq \mathbf{c}(u).$$

A vector $\tilde{\mathbf{c}}$ is an ϵ-absolute-approximation of the contribution vector $\mathbf{c} = \mathbf{cpr}(\alpha, v)$ if $\tilde{\mathbf{c}} \geq 0$ and, for all vertices u,

$$\mathbf{c}(u) - \epsilon \leq \tilde{\mathbf{c}}(u) \leq \mathbf{c}(u).$$

Clearly, an ϵ-approximation of $\mathbf{cpr}(\alpha, v)$ is an $(\epsilon \cdot \mathrm{pr}_\alpha(v))$-absolute-approximation of $\mathbf{cpr}(\alpha, v)$. In the algorithm below, we will focus on the computation of an ϵ-absolute-approximation of the contribution vector.

The *support* of a non-negative vector $\tilde{\mathbf{c}}$, denoted by $\mathrm{Supp}(\tilde{\mathbf{c}})$, is the set of all vertices whose entries in $\tilde{\mathbf{c}}$ are strictly positive. The vector $\mathbf{c}$ has a canonical ϵ-absolute-approximation. Let $\bar{\mathbf{c}}$ denote the vector

$$\bar{\mathbf{c}}(u) = \begin{cases} \mathbf{c}(u) \text{ if } \mathbf{c}(u) > \epsilon \\ 0 \quad \text{otherwise} \end{cases}.$$

Clearly, $\bar{\mathbf{c}}$ is the ϵ-absolute-approximation of $\mathbf{c}$ with the smallest support. Moreover, $\|\bar{\mathbf{c}}\|_1 \leq \|\mathbf{c}\|_1$ and thus, $|\mathrm{Supp}(\bar{\mathbf{c}})| \leq \|\mathbf{c}\|_1/\epsilon$. Our local algorithm attempts to find an approximation $\tilde{\mathbf{c}}$ of $\mathbf{c}$ which has a similar support structure to that of $\bar{\mathbf{c}}$.

3.1   High Level Idea of the Local Algorithm

It is well known that for each α, the personalized PageRank vector which satisfies Equation 2 also satisfies

$$\mathbf{ppr}(\alpha, u) = \alpha \sum_{t=0}^{\infty} (1 - \alpha)^t \cdot \left(\mathbf{e}_u M^t \right). \tag{3}$$

The contribution of u to v can then be written in the following way.

$$\mathrm{ppr}_\alpha(u \to v) = \langle \mathbf{ppr}(\alpha, u), \mathbf{e}_v \rangle \tag{4}$$

$$= \left\langle \alpha \sum_{t=0}^{\infty} (1-\alpha)^t (\mathbf{e}_u M^t), \mathbf{e}_v \right\rangle \tag{5}$$

$$= \left\langle \mathbf{e}_u, \alpha \sum_{t=0}^{\infty} (1-\alpha)^t (\mathbf{e}_v M^T)^t \right\rangle. \tag{6}$$

The standard way to compute the contribution of u to v is based on Equation 5. We refer to this approach as the time-forward calculation of $\mathrm{ppr}_\alpha(u \to v)$. Recall that $\mathbf{e}_u M^t$ is the t-step random walk distribution starting from u. In the time-forward calculation, we emulate the random walk from u step by step and add up the walk distributions scaled by the power sequence of $(1-\alpha)^t$. Without knowing in advance which vertices u make large contributions to v, one may have to perform the time-forward calculation of $\mathbf{ppr}(\alpha, u)$ for many vertices u to obtain a good approximation of $\mathbf{cpr}(\alpha, v)$.

To overcome this difficulty, we can directly calculate $\mathbf{cpr}(\alpha, v)$ in the manner suggested by Equation 6. This equation implies that

$$\mathbf{cpr}(\alpha, v) = \alpha \sum_{t=0}^{\infty} (1-\alpha)^t \cdot \left(\mathbf{e}_v (M^T)^t \right). \tag{7}$$

Thus, the contribution vector can be computed by starting with $\mathbf{e}_v$, iteratively computing $\mathbf{e}_v(M^T)^t$, and adding up the resulting vectors scaled by the power sequence of $(1-\alpha)^t$. Note that the matrix M^T is no longer a random walk matrix, since the sum of each row will not generally be equal to 1. Unlike the time-forward calculation, the direct calculation of $\mathbf{cpr}(\alpha, v)$ is no longer an emulation of the random walk starting from v. This fact complicates the error analysis of the next subsection.

The discussion above provides a way to directly compute $\mathbf{cpr}(\alpha, v)$, but our local algorithm will perform a different calculation. Instead of iteratively computing the vectors $\mathbf{e}_v(M^T)^t$, we adapt the technique of Jeh and Widom [14] for computing personalized PageRank to the task of computing contribution vectors. Using this method, we can compute the contribution vector in a decentralized way, and avoid spending computational effort manipulating small numerical values. This enables us to bound the running time required to obtain a fixed level of error.

Equation 7 also enables us to compute the vector of contributions to a specified subset S of vertices, which we define to be $\mathbf{cpr}(\alpha, S) = \sum_{v \in S} \mathbf{cpr}(\alpha, v)$. Let $\mathbf{e}_S = \sum_{v \in S} \mathbf{e}_v$. Then,

$$\mathbf{cpr}(\alpha, S) = \alpha \sum_{t=0}^{\infty} (1-\alpha)^t \cdot \left(\mathbf{e}_S (M^T)^t \right). \tag{8}$$

To further abuse notation, for any non-negative vector $\mathbf{s}$, we define

$$\mathbf{cpr}(\alpha, \mathbf{s}) = \alpha \sum_{t=0}^{\infty} (1-\alpha)^t \cdot \left(\mathbf{s} (M^T)^t \right). \tag{9}$$

3.2 The Local Algorithm and Its Analysis

The theorem below describes our algorithm `ApproxContributions` for computing an ϵ-absolute-approximation of the contribution vector of a target vertex v.
We give an upper bound on the number of vertices examined by the algorithm
that depends on $\mathrm{pr}_\alpha(v)$, ϵ, and α, but is otherwise independent of the number of
vertices in the graph. The algorithm performs a sequence of operations, which
we call pushback operations. Each pushback operation is performed on a single
vertex of the graph, and requires time proportional to the in-degree of that vertex. We place an upper bound on the *number* of pushback operations performed
by the algorithm, rather than the total running time of the algorithm. The total running time of the algorithm depends on the in-degrees of the sequence
of vertices on which the pushback operations were performed. The number of
pushback operations is an upper bound on the number of vertices in the support
of the resulting approximate contribution vector.

Theorem 1. *The algorithm* `ApproxContributions`$(v, \alpha, \epsilon, \mathbf{p}_{max})$ *has the following properties. The input is a vertex v, two constants α and ϵ in the interval
$(0, 1]$, and a real number $\mathbf{p}_{max}$. The algorithm computes a vector $\tilde{\mathbf{c}}$ such that
$0 \le \tilde{\mathbf{c}} \le \mathbf{c}$, and either*

1. $\tilde{\mathbf{c}}$ is an ϵ-absolute approximation of $\mathbf{cpr}(\alpha, v)$, or
2. $\|\tilde{\mathbf{c}}\|_1 \ge \mathbf{p}_{max}$.

*The number of pushback operations P performed by the algorithm satisfies the
following bound,*

$$P \le \frac{\min\left(\mathrm{pr}_\alpha(v), \mathbf{p}_{max}\right)}{\alpha\epsilon} + 1.$$

The proof of Theorem 1 is based on a series of facts which we describe below.
The starting point is the following observation, which is easy to verify from
Equation 9. For any vector $\mathbf{s}$,

$$\mathbf{cpr}(\alpha, \mathbf{s})M^T = \mathbf{cpr}(\alpha, \mathbf{s}M^T). \tag{10}$$

We can further derive the following equation,

$$\mathbf{cpr}(\alpha, \mathbf{s}) = \alpha\mathbf{s} + (1 - \alpha) \cdot \mathbf{cpr}(\alpha, \mathbf{s})M^T$$
$$= \alpha\mathbf{s} + (1 - \alpha) \cdot \mathbf{cpr}(\alpha, \mathbf{s}M^T). \tag{11}$$

This is the transposed version of the equation that was used Jeh and Widom
to compute approximate personalized PageRank vectors [14]. Very naturally, we
will use it to compute approximate contribution vectors.

The algorithm `ApproxContributions`$(v, \alpha, \epsilon, \mathbf{p}_{max})$ maintains a pair of vectors $\mathbf{p}$ and $\mathbf{r}$ with nonnegative entries, starting with the trivial approximation
$\mathbf{p} = \mathbf{0}$ and $r = \mathbf{e}_v$, and applies a series of pushback operations that increase
$\|\mathbf{p}\|_1$ while maintaining the invariant $\mathbf{p} + \mathbf{cpr}(\alpha, \mathbf{r}) = \mathbf{cpr}(\alpha, v)$. Each pushback
operation picks a single vertex u, moves an α fraction of the mass at $\mathbf{r}(u)$ to

$\mathbf{p}(u)$, and then modifies the vector $\mathbf{r}$ by replacing $\mathbf{r}(u)\mathbf{e}_u$ with $(1-\alpha)\mathbf{r}(u)\mathbf{e}_u M^T$. Note that $\|\mathbf{r}\|_1$ may increase or decrease during this operation. We will define the pushback operation more formally below, and then verify that each pushback operation does indeed maintain the invariant.

---

pushback (u):
Let $\mathbf{p}' = \mathbf{p}$ and $\mathbf{r}' = \mathbf{r}$, except for these changes:

1. $\mathbf{p}'(u) = \mathbf{p}(u) + \alpha \mathbf{r}(u)$.
2. $\mathbf{r}'(u) = 0$.
3. For each vertex w such that $w \to u$:
   $\mathbf{r}'(w) = \mathbf{r}(w) + (1-\alpha)\mathbf{r}(u)/d_{out}(w)$.

Lemma 1 (Invariant). *Let $\mathbf{p}'$ and $\mathbf{r}'$ be the result of performing* pushback(u) *on $\mathbf{p}$ and $\mathbf{r}$. If $\mathbf{p}$ and $\mathbf{r}$ satisfy the invariant $\mathbf{p} + \mathbf{cpr}(\alpha, \mathbf{r}) = \mathbf{cpr}(\alpha, v)$, then $\mathbf{p}'$ and $\mathbf{r}'$ satisfy the invariant $\mathbf{p}' + \mathbf{cpr}(\alpha, \mathbf{r}') = \mathbf{cpr}(\alpha, v)$.*

Proof. After the pushback operation, we have, in vector notation,

$$\mathbf{p}' = \mathbf{p} + \alpha \mathbf{r}(u)\mathbf{e}_u.$$
$$\mathbf{r}' = \mathbf{r} - \mathbf{r}(u)\mathbf{e}_u + (1-\alpha)\mathbf{r}(u)\mathbf{e}_u M^T.$$

We will apply equation (11) to $\mathbf{r}(u)\mathbf{e}_u$ to show that $\mathbf{p} + \mathbf{cpr}(\alpha, \mathbf{r}) = p' + \mathbf{cpr}(\alpha, \mathbf{r}')$.

$$
\begin{aligned}
\mathbf{cpr}(\alpha, \mathbf{r}) &= \mathbf{cpr}(\alpha, \mathbf{r} - \mathbf{r}(u)\mathbf{e}_u) + \mathbf{cpr}(\alpha, \mathbf{r}(u)\mathbf{e}_u) \\
&= \mathbf{cpr}(\alpha, \mathbf{r} - \mathbf{r}(u)\mathbf{e}_u) + \alpha \mathbf{r}(u)\mathbf{e}_u + \mathbf{cpr}(\alpha, (1-\alpha)\mathbf{r}(u)\mathbf{e}_u M^T) \\
&= \mathbf{cpr}(\alpha, \mathbf{r} - \mathbf{r}(u)\mathbf{e}_u + (1-\alpha)\mathbf{r}(u)\mathbf{e}_u M^T) + \alpha \mathbf{r}(u)\mathbf{e}_u \\
&= \mathbf{cpr}(\alpha, \mathbf{r}') + \mathbf{p}' - \mathbf{p}.
\end{aligned}
$$

During each pushback operation, the quantity $\|\mathbf{p}\|_1$ increases by $\alpha \mathbf{r}(u)$. The quantity $\|\mathbf{p}\|_1$ can never exceed $\|\mathbf{cpr}(\alpha, v)\|_1$, which is equal to $\mathrm{pr}_\alpha(v)$. By performing pushback operations only on vertices where $\mathbf{r}(u) \geq \epsilon$, we can ensure that $\|\mathbf{p}\|_1$ increases by a significant amount at each step, which allows us to bound the number of pushes required to compute an ϵ-absolute-approximation of the contribution vector. This is the idea behind the algorithm ApproxContributions.

ApproxContributions$(v, \alpha, \epsilon, \mathbf{p}_{max})$:

1. Let $\mathbf{p} = 0$, and $\mathbf{r} = \mathbf{e}_v$.
2. While $\mathbf{r}(u) > \epsilon$ for some vertex u:
   (a) Pick any vertex u where $\mathbf{r}(u) \geq \epsilon$.
   (b) Apply pushback (u).
   (c) If $\|\mathbf{p}\|_1 \geq \mathbf{p}_{max}$, halt and output $\tilde{\mathbf{c}} = \mathbf{p}$.
3. Output $\tilde{\mathbf{c}} = \mathbf{p}$.

This algorithm can be implemented by maintaining a queue containing those vertices u satisfying $\mathbf{r}(u) \geq \epsilon$. Initially, v is the only vertex in the queue. At each step, we take the first vertex u in the queue, remove it from the queue, and perform a pushback operation from that vertex. If the pushback operation raises the value of $\mathbf{r}(x)$ above ϵ for some in-neighbor x of u, then x is added to the back of the queue. This continues until the queue is empty, at which point all vertices satisfy $\mathbf{r}(u) < \epsilon$, or until $\|\mathbf{p}\|_1 \geq \mathbf{p}_{max}$. We now show that this algorithm has the properties promised in Theorem 1.

*Proof (***Proof of Theorem 1***).* Let T be the total number of push operations performed by the algorithm, and let $\mathbf{p}_t$ and $\mathbf{r}_t$ be the states of the vectors $\mathbf{p}$ and $\mathbf{r}$ after t pushes. The initial setting of $\mathbf{p}_0 = 0$ and $\mathbf{r}_0 = \mathbf{e}_v$ satisfies the invariant $\mathbf{p}_t + \mathbf{cpr}(\alpha, \mathbf{r}_t) = \mathbf{cpr}(\alpha, v)$, which is maintained throughout the algorithm. Since $\mathbf{r}_t$ is nonnegative at each step, the error term $\mathbf{cpr}(\alpha, \mathbf{r}_t)$ is also nonnegative, so we have $\mathbf{cpr}(\alpha, v) - \mathbf{p}_t \geq 0$. In particular, this implies $\|\mathbf{p}_t\|_1 \leq \|\mathbf{cpr}(\alpha, v)\|_1 = \mathrm{pr}_\alpha(v)$.

Let $\tilde{\mathbf{c}} = \mathbf{p}_T$ be the vector output by the algorithm. When the algorithm terminates, we must have either $\|\tilde{\mathbf{c}}\|_1 \geq \mathbf{p}_{max}$ or $\|\mathbf{r}_T\|_\infty \leq \epsilon$. In the latter case, the following calculation shows that $\tilde{\mathbf{c}}$ is an ϵ-absolute-approximation of $\mathbf{cpr}(\alpha, v)$.

$$\|\mathbf{cpr}(\alpha, v) - \tilde{\mathbf{c}}\|_\infty = \|\mathbf{cpr}(\alpha, \mathbf{r}_T)\|_\infty$$
$$\leq \|\mathbf{r}_T\|_\infty$$
$$\leq \epsilon.$$

The fact that $\|\mathbf{cpr}(\alpha, \mathbf{r}_T)\|_\infty \leq \|\mathbf{r}_T\|_\infty$ holds because $\mathbf{r}_T$ is nonnegative and each row of M sums to 1.

The vector $\mathbf{p}_{T-1}$ must have satisfied $\|\mathbf{p}_{T-1}\|_1 < \mathbf{p}_{max}$, since the algorithm decided to push one more time. We have already observed that $\|\mathbf{p}_{T-1}\|_1 \leq \mathrm{pr}_\alpha(v)$. Each push operation increased $\|\mathbf{p}\|_1$ by at least $\alpha\epsilon$, so we have

$$\alpha\epsilon(T - 1) \leq \|\mathbf{p}_{T-1}\|_1 \leq \min\left(\|\mathbf{cpr}(\alpha, v)\|_1, \mathbf{p}_{max}\right).$$

This gives the desired bound on T.

It is possible to perform a pushback operation on the vertex u, and to perform the necessary queue updates, in time proportional to $d_{in}(u)$. Therefore, the running time of the algorithm is proportional to the sum over all pushback operations of the in-degree of the pushed vertex.

We can compute an ϵ-approximation of $\mathbf{cpr}(\alpha, v)$, provided that $\mathrm{pr}_\alpha(v)$ is known, by calling the algorithm `ApproxContributions`$(v, \alpha, \epsilon \cdot \mathrm{pr}_\alpha(v), \mathrm{pr}_\alpha(v))$.

Corollary 1 (ϵ-Approximation of contribution vectors). *Given* $\mathrm{pr}_\alpha(v)$, *an ϵ-approximation of* $\mathbf{cpr}(\alpha, v)$, *can be computed with* $\frac{1}{\alpha\epsilon} + 1$ *pushback operations.*

We also observe that, using Equation 8, our algorithm can be easily adapted to compute an ϵ-absolute-approximation and ϵ-approximation of $\mathbf{cpr}(\alpha, S)$ for a group S of vertices, with a similar bound on the number of pushback operations.

3.3   The Support of the Approximate Contribution Vector

The number of vertices in the support of the ϵ-approximate contribution vector $\tilde{c}$ is upper bounded by the number of pushback operations used to compute it, which is at most $\frac{1}{\alpha\epsilon} + 1$. In this section we give a stronger upper bound on the size of the support. To do this, we need to modify the pushback operation slightly. Instead of moving all the mass from $\mathbf{r}(u)$ during the pushback operation, we move all but $\epsilon/2$ units of mass, and leave $\epsilon/2$ units on $\mathbf{r}(u)$. This increases the running time bound for the algorithm by a factor of 2, but ensures that $\mathbf{r}(x) \geq \epsilon/2$ at each vertex in $\mathrm{Supp}(\tilde{c})$. We use this fact to give a family of bounds on the size of $\mathrm{Supp}(\tilde{c})$.

We will abuse our notation a bit by defining the following,

$$\mathrm{pr}_\alpha(\mathbf{x} \to \mathbf{y}) = \langle \mathbf{x}M_\alpha , \mathbf{y} \rangle,$$

where $M_\alpha = \mathbf{PRM}_\alpha$ is the PageRank matrix. In particular, $\mathrm{pr}_\alpha(\mathbf{x} \to \mathbf{e}_S)$ is the amount probability from the PageRank vector with starting distribution $\mathbf{x}$ on the set S.

Proposition 1. *Let $\tilde{c}$ be the ϵ-approximate contribution vector for v computed by the modified algorithm described above, and let $S = \mathrm{Supp}(\tilde{c})$. For any nonnegative vector $\mathbf{z}$, we have the following upper bound on S,*

$$\mathrm{pr}_\alpha(\mathbf{z} \to \mathbf{e}_S) \leq \frac{2}{\epsilon}\mathrm{pr}_\alpha(\mathbf{z} \to \mathbf{e}_v).$$

Proof. Note that $\mathbf{ppr}(\alpha, v) = \mathbf{e}_v M_\alpha$ and $\mathbf{cpr}(\alpha, v) = \mathbf{e}_v M_\alpha^T$. We know that $\mathbf{cpr}(\alpha, \mathbf{r}) \leq \mathbf{cpr}(\alpha, \mathbf{e}_v)$, which can also be written $\mathbf{r}M_\alpha^T \leq \mathbf{e}_v M_\alpha^T$. Let $S = \mathrm{Supp}(\tilde{c})$ and recall that $\mathbf{r}(x) \geq \epsilon/2$ for any vertex $x \in S$. Then,

$$\langle \mathbf{z}M_\alpha , \mathbf{e}_v \rangle = \langle \mathbf{z}, \mathbf{e}_v M_\alpha^T \rangle \geq \langle \mathbf{z}, \mathbf{r}M_\alpha^T \rangle = \langle \mathbf{z}M_\alpha , \mathbf{r} \rangle \geq (\epsilon/2)\langle \mathbf{z}M_\alpha , \mathbf{e}_S \rangle.$$

In the second step we needed $\mathbf{z}$ to be nonnegative, and in the last step we needed $\mathbf{z}M_\alpha$ to be nonnegative, which is true whenever $\mathbf{z}$ is nonnegative.

In words, this proposition states that for any starting vector $\mathbf{z}$, the amount of probability from the PageRank vector $\mathbf{ppr}(\alpha, \mathbf{z})$ on the set $S = \mathrm{Supp}(\tilde{c})$ is at most $2/\epsilon$ times the amount on the vertex v. If we let $\mathbf{z} = \mathbf{e}_V$, then we obtain a bound on the amount of global PageRank on the set S,

$$\mathrm{pr}_\alpha(S) \leq \frac{2}{\epsilon}\mathrm{pr}_\alpha(v).$$

To see that this bound is at least as strong as what we knew before, recall that the PageRank of any given vertex is at least α. If we make the pessimistic assumption that $\mathrm{pr}_\alpha(u) = \alpha$ for each $u \in \mathrm{Supp}(\tilde{c})$, then the bound we have just proved reduces to our earlier bound on the number of pushback operations,

$$|\mathrm{Supp}(\tilde{c})| \leq 2\mathrm{pr}_\alpha(v)/\alpha\epsilon.$$

4   Computing Supporting Sets

In this section, we use our local algorithm for approximating contribution vectors to compute approximate supporting sets, sets of vertices that contribute significantly to the PageRank of a target vertex. There are several natural notions of supporting sets, which we define below. For a vertex v, let π_v be the permutation that orders the entries $\mathbf{cpr}(\alpha, v)$ from the largest to the smallest. Ties may be broken arbitrarily.

- **top k contributors:** the first k pages of π_v.
- **δ-significant contributors:** $\{u \mid \mathrm{ppr}_\alpha(u \to v) > \delta\}$.
- **ρ-supporting set:** a set S of pages such that

$$\mathrm{ppr}_\alpha(S \to v) \geq \rho \cdot \mathrm{pr}_\alpha(v).$$

In addition, let $k_\rho(v)$ be the smallest integer such that

$$\mathrm{ppr}_\alpha(\pi_v(1 : k_\rho(v)) \to v) \geq \rho \cdot \mathrm{pr}_\alpha(v).$$

Clearly the set of the first $k_\rho(v)$ pages of π_v is the minimum size ρ-supporting set for v. Also, we define $\rho_k(v) = \mathrm{ppr}_\alpha(\pi_v(1 : k) \to v)/\mathrm{pr}_\alpha(v)$ to be the fraction of v's PageRank contributed by its top k contributors.

4.1   Approximating Supporting Sets

Without precisely computing $\mathbf{cpr}(\alpha, v)$ it might be impossible to identify supporting sets exactly, so we consider approximate supporting sets. For a precision parameter ϵ, we define the following.

- **ϵ-precise top k contributors:** a set of k pages that contains all pages whose contribution to v is at least $\mathrm{ppr}_\alpha(\pi_v(k) \to v) + \epsilon \cdot \mathrm{pr}_\alpha(v)$, but no page with contribution to v less than $\mathrm{ppr}_\alpha(\pi_v(k) \to v) - \epsilon \cdot \mathrm{pr}_\alpha(v)$.
- **ϵ-precise δ-significant contributors:** a set that contains the set of δ-significant contributors and is contained in the set of $(\delta - \epsilon)$-significant contributors.

The results in the remainder of this section assume that $\mathrm{pr}_\alpha(v)$ is known.

Theorem 2. *An ϵ-precise set of top k contributors of a vertex v can be found by performing $1/\alpha\epsilon + 1$* pushback *operations.*

Proof. Call $\tilde{\mathbf{c}} = \texttt{ApproxContributions}(v, \alpha, \epsilon \cdot \mathrm{pr}_\alpha(v), \mathrm{pr}_\alpha(v))$. Let $C = \mathrm{Supp}(\tilde{\mathbf{c}})$. If $|C| > k$, then return the vertices with the top k entries in $\tilde{\mathbf{c}}$; otherwise, return C together with $k - \mathrm{Supp}(\tilde{\mathbf{c}})$ arbitrarily chosen vertices not in C. Consider a page u with $\mathbf{cpr}(u, v) \geq \mathbf{cpr}(\pi_v(k), v) + \epsilon \cdot \mathrm{pr}_\alpha(v)$. Clearly $u \in C$ because $\tilde{\mathbf{c}}(u) \geq \mathbf{cpr}(\pi_v(k), v)$, implying $\tilde{\mathbf{c}}(u)$ is among the top k entries in $\tilde{\mathbf{c}}$. On the other hand, $\tilde{\mathbf{c}}(\pi_v(j))$ is at least $\mathbf{cpr}(\pi_v(k), v) - \epsilon \cdot \mathrm{pr}_\alpha(v)$ for all $j \in [1 : k]$. Thus, each of the vertices with the top k entries in $\tilde{\mathbf{c}}$ must contribute at least $\mathbf{cpr}(\pi_v(k), v) - \epsilon \cdot \mathrm{pr}_\alpha(v)$ to v.

Theorem 3. *An ϵ-precise δ-significant contributing set of a vertex v can be found by performing $1/\alpha\epsilon + 1$ pushback operations.*

Proof. Call $\tilde{\mathbf{c}} = $ ApproxContributions$(v, \alpha, \epsilon \cdot \mathrm{pr}_\alpha(v), \mathrm{pr}_\alpha(v))$ and return the vertices whose entries in $\tilde{\mathbf{c}}$ are at least $(\delta - \epsilon) \cdot \mathrm{pr}_\alpha(v)$. Clearly, the set contains the δ-contributing set of v and is contained in the $(\delta - \epsilon)$-supporting set of v. Moreover, the number of pages not in the δ-supporting set that are included is at most $1/(\delta - \epsilon)$.

In the remainder of this section, we consider the computation of approximate ρ-supporting sets. We give two different algorithms, one for finding a supporting set on a fixed number of vertices with the largest contribution possible, and one for finding a supporting set with a fixed contribution on as few vertices as possible.

Theorem 4. *Given a vertex v and an integer k, a set of k vertices that is a $(\rho_k - \epsilon)$-supporting set for v can be found by performing $k/\alpha\epsilon + 1$ pushback operations.*

Proof. Compute $\tilde{\mathbf{c}} = $ ApproxContributions$(v, \alpha, \epsilon\,\mathrm{pr}_\alpha(v)/k, \mathrm{pr}_\alpha(v))$. Let S_k be the set of k top contributors to v, which are the k vertices with the highest values in $\mathbf{c}$, and let $\tilde{S}_k$ be the set of k vertices with the highest values in $\tilde{\mathbf{c}}$. The set $\tilde{S}_k$ meets the requirements of the theorem, since we have

$$\tilde{\mathbf{c}}(\tilde{S}_k) \geq \mathbf{c}(S_k) - k(\epsilon\,\mathrm{pr}_\alpha(v)/k)$$
$$\geq \rho_k \cdot \mathrm{pr}_\alpha(v) - \epsilon \cdot \mathrm{pr}_\alpha(v)$$
$$= \mathrm{pr}_\alpha(v)(\rho_k - \epsilon).$$

Theorem 5. *Assume we are given ρ but not k_ρ. A set of at most k_ρ vertices that is a $(\rho - \epsilon)$-supporting set for v can be found by performing $O(k_\rho \log k_\rho/\alpha\epsilon)$ pushback operations.*

Proof. The challenge here is that we do not know k_ρ, so we need to use a binary search procedure to find a proxy for k_ρ. We will proceed in two phases. In the first phase, we guess a value of k, starting with $k = 1$, and compute $\tilde{\mathbf{c}} = $ ApproxContributions$(v, \alpha, \epsilon \cdot \mathrm{pr}_\alpha(v)/k, \mathrm{pr}_\alpha(v))$. As in Theorem 4, let $\tilde{S}_k$ be the set of k vertices with the highest values in $\tilde{\mathbf{c}}$, which we know satisfies $\tilde{\mathbf{c}}(\tilde{S}_k) \geq (\rho_k - \epsilon)$. If we observe that $\tilde{\mathbf{c}}(\tilde{S}_k) < (\rho - \epsilon)$, then we double k and repeat the procedure. If we observe that $\tilde{\mathbf{c}}(\tilde{S}_k) \geq (\rho - \epsilon)$, then we halt and proceed to the second phase, and set k_1 to be the value of k for which this happens. We must have $k_1 \leq 2k_\rho$, since we are guaranteed to halt if $k \geq k_\rho$.

Let $k_0 = k_1/2$ be the value of k from the step before the first phase halted. In the second phase, we perform binary search within the interval $[k_0, k_1]$ to find the smallest integer k_{min} for which $\tilde{\mathbf{c}}(\tilde{S}_{k_{min}}) \geq (\rho - \epsilon)$, which must satisfy $k_{min} \leq k_\rho$. We output $\tilde{S}_{k_{min}}$.

Each time we call the subroutine $\tilde{\mathbf{c}} = $ ApproxContributions$(v, \alpha, \epsilon\,\mathrm{pr}_\alpha(v)/k, \mathrm{pr}_\alpha(v))$, it requires $k/\alpha\epsilon + 1$ push operations. In the first phase we call this subroutine with a sequence of k values that double from 1 up to at most $2k_\rho$, so the number of push operations performed is $O(k_\rho/\alpha\epsilon + \log k_\rho)$. In the second phase, the binary search makes at most $\log k_\rho$ calls to the subroutine, with k set to at most $2k_\rho$ in each step, so the number of push operations performed is $O(k_\rho \log k_\rho/\alpha\epsilon + \log k_\rho)$. The total number of push operations performed in both phases is $O(k_\rho \log k_\rho/\alpha\epsilon)$.

4.2   Local Estimation of PageRank

Up to this point, we have assumed when computing the supporting set of a vertex that its PageRank is known. We now consider how to apply our approximate contribution algorithm when nothing is known about the PageRank of the target vertex. In particular, we consider the problem of computing a lower bound on the PageRank of a vertex using local computation.

A natural lower bound on the PageRank $\mathrm{pr}_\alpha(v)$ is provided by the contribution to v of its top k contributors, $p_k = \mathbf{cpr}(\pi_v(1:k), v)$. The theorem below shows we can efficiently certify that $\mathrm{pr}_\alpha(v)$ is approximately as large as p_k without prior knowledge of $\mathrm{pr}_\alpha(v)$ or p_k. This should be contrasted with the algorithms from the previous section, for which we needed to know the value $\mathrm{pr}_\alpha(v)$ in order to set ϵ to obtain the stated running times.

Theorem 6. *Given k and δ, we can compute a real number p such that*

$$p_k(1+\delta)^{-2} \leq p \leq \mathrm{pr}_\alpha(v),$$

where $p_k = \mathbf{cpr}(\pi_v(1:k), v)$, by performing $10k\log(k/\alpha\delta)/\alpha$ pushback *operations.*

Proof. Fix k and δ, choose a value of p, and compute $\tilde{c} = $ ApproxContributions (v, α, ϵ, p) with $\epsilon = \delta p/k$. The number of pushback operations performed is at most

$$1 + p/\alpha\epsilon = 1 + p/\alpha(\delta p/k) = 1 + 10k/\alpha.$$

When the algorithm halts, we either have $\|\tilde{c}\|_1 \geq p$, in which case we have certified that $\mathrm{pr}_\alpha(v) \geq p$, or else we have $\|\tilde{c} - \mathbf{cpr}(\alpha, v)\|_\infty \leq \delta p/k$, in which case we have certified that $p_k \leq (1+\delta)p$, by the following calculation:

$$p_k = \mathbf{cpr}(\pi_v(1:k), v) \leq \tilde{c}(\pi_v(1:k), v) + (\delta p/k)k \leq p + \delta p.$$

We now perform binary search over p in the range $[\alpha, k]$. Let p_{low} be the largest value of p for which we have certified that $\mathrm{pr}_\alpha(v) \geq p$, and let p_{high} be the smallest value of p for which we have certified that $p_k \leq (1+\delta)p$. We perform binary search until $p_{high} \leq p_{low}(1+\delta)$, which requires at most $\log(k/\alpha\delta)$ steps. Then, p_{low} has the property described in the theorem,

$$\mathrm{pr}_\alpha(v) \geq p_{low} \geq p_{high}(1+\delta)^{-1} \geq p_k(1+\delta)^{-2}.$$

The total number of pushback operations performed during the calls to ApproxContributions during the binary search is at most $10k\log(k/\alpha\delta)/\alpha$.

5 Final Remarks

5.1 Improving the Dependency on In-Degrees

In our performance analysis, we give a bound of $\mathrm{pr}_\alpha(v)/(\alpha\epsilon) + 1$ on the total number of **pushback** operations performed by our algorithm. In a pushback at a vertex u, we update the entry for u in the vector **p** as well the as entries in **r** for all vertices that point to u. As a result, the overall time complexity of our algorithm is proportional to the sum of the in-degrees of the sequence of vertices that we pushback from. A possible direction for future research is to devise an algorithm whose running time can be bounded in terms of the total in-degree of the supporting set that the algorithm attempts to approximate. This type of bound would offer stronger control over the running time than the result obtained in this paper, where the number of pushback operations operations is bounded in terms of the number of vertices in the supporting set, but the running time depends on the in-degrees of the vertices from which the sequence of push operations is performed.

5.2 Computing Contribution Vectors Via the Time-Reverse Chain

As noted earlier, the matrix M^T in the formula of Equation 7 may not be Markov. It is natural ask whether the time-reverse Markov chain of the random walk matrix M may be used to compute the contribution vector for a vertex v, and, if so, whether this method is efficient.

For the following discussion, we assume that M has a unique stationary distribution, which will not be true for general directed graphs. Recall that,

Definition 2 (Time-reverse chain). *Given a Markov chain M with transition probability m_{ij}, and stationary distribution π, the time-reverse chain is the Markov chain R with transition probability $r_{ij} = \pi(j)m_{ji}/\pi(i)$.*

In other words, let Π be the matrix whose (i,j)th entry is $\pi(j)/\pi(i)$, then $R = \Pi \cdot {*}M^T$, where the operation $\cdot{*}$ is the component-wise multiplication of two matrices. The time-reverse chain has the following properties.

- R has the same stationary distribution as M,
- for all i, k, and t, consider the t-step random walk starting from i in M and k in R, then

$$\left\langle\, \mathbf{e}_i M^t \,,\, \mathbf{e}_k \,\right\rangle = \left(\frac{\pi(k)}{\pi(i)} \right) \left\langle\, \mathbf{e}_k R^t \,,\, \mathbf{e}_i \,\right\rangle \tag{12}$$

Recall $\left\langle\, \mathbf{e}_i M^t \,,\, \mathbf{e}_k \,\right\rangle$ is equal to the probability that k is the vertex reached by a t-step random walk from i. Let $\mathrm{ppr}_\alpha^M(u \to v)$ denote the personalized PageRank contribution from u to v in a Markov chain M.

Theorem 7. *Suppose a Markov chain M has a stationary distribution π and R is its time-reverse chain. Then*

$$\mathrm{ppr}_\alpha^M(u \to v) = \left(\frac{\pi(v)}{\pi(u)}\right) \mathrm{ppr}_\alpha^R(v \to u). \tag{13}$$

Proof. The result follows from Equations 5 and 12.

Thus, if the stationary distribution exists, we can in principle compute the contribution vector of M by computing the personalized PageRank vector for v in the time-reverse chain. We argue that the method we presented in Section 3 is preferable to the time-reverse Markov chain method for the following reasons. Our method does not require that M has a stationary distribution. Computing a personalized PageRank vector in the time-reverse Markov chain requires that we first compute the stationary distribution π of M, which may be computationally expensive. Perhaps most important is the difference in the error analysis. If the stationary distribution exists, one can compute an ϵ-approximate contribution vector by computing a personalized PageRank vector in R for which the error at each vertex i is at most $\epsilon\pi(i)$. If $\pi(i)$ is extremely small at some vertices, and it may be exponentially small in the number of vertices in the graph, this will require a large amount of computation.

We prefer the method presented in Section 3 to the time-reverse method for most graphs that are likely to be encountered in practice. However, there are special cases where the time-reverse method will be efficient. In particular, if the Markov chain has a stationary distribution that is nearly proportional to the in-degrees of the vertices, as it would be in an undirected graph, then computing a personalized PageRank vector in the time-reverse chain is an efficient way to compute a contribution vector.

References

1. Andersen, R., Borgs, C., Chayes, J., Hopcroft, J., Jain, K., Mirrokni, V., Teng, S.: Experimental evaluation of locally computable link-spam features (submitted, 2007)
2. Andersen, R., Chung, F., Lang, K.: Local graph partitioning using pagerank vectors. In: FOCS 2006: Proceedings of the 47th Annual IEEE Symposium on Foundations of Computer Science, pp. 475–486. IEEE Computer Society, Washington, DC (2006)
3. Becchetti, L., Castillo, C., Donato, D., Leonardi, S., Baeza-Yates, R.: Link-based characterization and detection of web spam (2006)
4. Benczúr, A.A., Csalogány, K., Sarlós, T., Uher, M.: Spamrank - fully automatic link spam detection. In: First International Workshop on Adversarial Information Retrieval on the Web (2005)
5. Berkhin, P.: Bookmark-coloring algorithm for personalized pagerank computing. Internet Math. 3(1), 41–62 (2006)
6. Brin, S., Page, L.: The anatomy of a large-scale hypertextual Web search engine. Computer Networks and ISDN Systems 30(1-7), 107–117 (1998)

7. Chen, Y., Gan, Q., Suel, T.: Local methods for estimating pagerank values. In: Proc. of CIKM, pp. 381–389 (2004)
8. Fetterly, D., Manasse, M., Najork, M.: Spam, damn spam, and statistics: using statistical analysis to locate spam web pages. In: WebDB 2004: Proceedings of the 7th International Workshop on the Web and Databases, pp. 1–6. ACM Press, New York (2004)
9. Fogaras, D., Racz, B.: Towards scaling fully personalized pagerank. In: Leonardi, S. (ed.) WAW 2004. LNCS, vol. 3243, pp. 105–117. Springer, Heidelberg (2004)
10. Gyöngyi, Z., Berkhin, P., Garcia-Molina, H., Pedersen, J.: Link spam detection based on mass estimation. In: Proceedings of the 32nd International Conference on Very Large Databases, ACM, New York (2006)
11. Gyöngyi, Z., Garcia-Molina, H., Pedersen, J.: Combating web spam with trustrank. In: VLDB, pp. 576–587 (2004)
12. Gyöngyi, Z., Garcia-Molina, H., Pedersen, J.: Web content categorization using link information. Technical report, Stanford University (2006)
13. Haveliwala, T.H.: Topic-sensitive pagerank: A context-sensitive ranking algorithm for web search. IEEE Trans. Knowl. Data Eng. 15(4), 784–796 (2003)
14. Jeh, G., Widom, J.: Scaling personalized web search. In: WWW 2003. Proceedings of the 12th World Wide Web Conference, pp. 271–279 (2003)
15. Mishne, G., Carmel, D.: Blocking blog spam with language model disagreement (2005)
16. Naor, M., Stockmeyer, L.: What can be computed locally? SIAM J. Comput. 24(6), 1259–1277 (1995)
17. Ntoulas, A., Najork, M., Manasse, M., Fetterly, D.: Detecting spam web pages through content analysis. In: WWW 2006: Proceedings of the 15th international conference on World Wide Web, pp. 83–92. ACM Press, New York (2006)
18. Raj, R., Krishnan, V.: Web spam detection with anti-trust rank. In: Proc. of the 2nd International Worshop on Adversarial Information Retreival on the Web, pp. 381–389 (2006)
19. Sarlós, T., Benczúr, A.A., Csalogány, K., Fogaras, D.: To randomize or not to randomize: space optimal summaries for hyperlink analysis. In: WWW, pp. 297–306 (2006)
20. Spielman, D.A., Teng, S.-H.: Nearly-linear time algorithms for graph partitioning, graph sparsification, and solving linear systems. In: ACM STOC-04, pp. 81–90. ACM Press, New York (2004)

Local Partitioning for Directed Graphs Using PageRank

Reid Andersen[1], Fan Chung[2], and Kevin Lang[3]

[1] Microsoft Research, Redmond WA 98052
reidan@microsoft.com
[2] University of California, San Diego, La Jolla CA 92093-0112
fan@ucsd.edu
[3] Yahoo! Research, Santa Clara CA 95054
langk@yahoo-inc.com

Abstract. A local partitioning algorithm finds a set with small conductance near a specified seed vertex. In this paper, we present a generalization of a local partitioning algorithm for undirected graphs to strongly connected directed graphs. In particular, we prove that by computing a personalized PageRank vector in a directed graph, starting from a single seed vertex within a set S that has conductance at most α, and by performing a sweep over that vector, we can obtain a set of vertices S' with conductance $\Phi_M(S') = O(\sqrt{\alpha \log |S|})$. Here, the conductance function Φ_M is defined in terms of the stationary distribution of a random walk in the directed graph. In addition, we describe how this algorithm may be applied to the PageRank Markov chain of an arbitrary directed graph, which provides a way to partition directed graphs that are not strongly connected.

1 Introduction

In directed networks like the world wide web, it is critical to develop algorithms that utilize the additional information conveyed by the direction of the links. Algorithms for web crawling, web mining, and search ranking, all depend heavily on the directedness of the graph. For the problem of graph partitioning, it is extremely challenging to develop algorithms that effectively utilize the directed links.

Spectral algorithms for graph partitioning have natural obstacles for generalizations to directed graphs. Nonsymmetric matrices do not have a spectral decomposition, meaning there does not necessarily exist an orthonormal basis of eigenvectors. The stationary distribution for random walks on directed graphs is no longer determined by the degree sequences. In the earlier work of Fill [7] and Mihail [12], several generalizations for directed graphs were examined for regular graphs. Lovász and Simonovits [11] established a bound for the mixing rate of an asymmetric ergodic Markov chain in terms of its conductance. When applied to the Markov chain of a random walk in a strongly connected directed graph, their results can be used to identify a set of states of the Markov chain with small conductance. Algorithms for finding sparse cuts, based on linear and semidefinite programming and metric embeddings, have also been generalized to

A. Bonato and F.R.K. Chung (Eds.): WAW 2007, LNCS 4863, pp. 166–178, 2007.
© Springer-Verlag Berlin Heidelberg 2007

directed graphs [3,6]. A Cheeger inequality for directed graphs which relies on the eigenvalues of a normalized Laplacian for directed graphs can also be used to find cuts of small conductance [5].

This paper is concerned with a different type of partitioning algorithm, called a *local partitioning algorithm*. A local partitioning algorithm finds a set with small conductance near a specified seed vertex, and can produce such a cut by examining only a small portion of the input graph. In a recent paper, the authors introduced a local partitioning algorithm, for undirected graphs, that finds a cut with small conductance by performing a sweep over a personalized PageRank vector. Personalized PageRank traditionally has been applied and studied in directed web graphs, so it is natural to ask whether this local partitioning algorithm can be generalized to find sets with small conductance in a directed graph by sweeping over a personalized PageRank vector computed in a directed graph.

In this paper, we generalize the basic local partitioning results from [1] to strongly connected directed graphs. We prove that by computing a personalized PageRank vector in a directed graph, and sorting the vertices of the graph according to their probability in this vector divided by their probability in the stationary distribution, we can identify a set with small conductance, where the notion of conductance must be generalized appropriately. Directed graphs that arise in practice are typically not strongly connected, and this generalized local partitioning algorithm cannot be applied directly to such a graph. We address this problem by describing how our algorithm may be applied to the PageRank Markov chain of a directed graph, which is ergodic even when the underlying graph is not strongly connected. When applied to the PageRank Markov chain, the generalized local partitioning algorithm has a natural interpretation: we compute a personalized PageRank vector with a single starting vertex, and a global PageRank vector with a uniform starting vector, and sort the vertices of the graph according to the ratio of their entries in the personalized PageRank vector and global PageRank vector. We prove that by sorting the vertices of the graph according to this ratio, our algorithm finds a set with small conductance in the PageRank Markov chain. We also show that the required computation can be carried out efficiently.

The generalized local partitioning algorithm has advantages and disadvantages when compared to the undirected algorithm. One advantage is that our algorithm follows outlinks exclusively, and does not travel backwards over inlinks. This ensures that all the vertices in the resulting cut are reachable from the starting vertex, and is particularly useful in settings where outlinks are more easily accessible than inlinks. One disadvantage is that the appropriate generalization of conductance to directed graphs requires reweighting the edges of the graph according to the amount of probability moving over them in the stationary distribution π of a random walk, which is more complicated in a directed graph than in the undirected case. The generalized local partitioning algorithm is guaranteed to find a cut for which the total weight of outlinks crossing the cut is small, but this weight depends on π, and the cut may have a large number of outlinks with small weight.

Here is an outline of the paper. In the next section, we define the generalizations of the key ingredients of the local partitioning algorithm from [1] to
strongly connected directed graphs, including personalized PageRank, conductance, sweeps, and the Lováasz-Simonovits potential function. In the main section, we prove a generalization of our basic local partitioning results to strongly
connected directed graphs. We prove that that a sweep over a personalized
PageRank vector in the directed graph produces a set with small conductance.
In Section 6, we describe how to apply our algorithm to the PageRank matrix of
an arbitrary directed graph, which is always strongly connected. We will show
that our local algorithm can find sets with small conductance by computing personalized PageRank vectors in the original directed graph, provided we compute
two global PageRank vectors offline.

2   Preliminaries

Let G be a directed graph, consisting of a vertex set V and a set of directed
edges E, each of which is an ordered pair (u, v) of vertices from V. Let n be the
number of vertices, and m be the number of directed edges. We write $d_{out}(v)$ for
the out-degree of a vertex v.

The adjacency matrix $A = A(G)$ is the $n \times n$ matrix where $A_{i,j} = 1$ if and
only if there is a directed edge (v_i, v_j), given some fixed ordering $v_1, \ldots, v_n$ of the
vertices. The out-degree matrix $D = D(G)$ is the $n \times n$ diagonal matrix where
$D_{i,i} = d_{out}(v_i)$.

For a given directed graph, we will consider several different Markov chains.
For our purposes, a Markov chain M is the matrix of a random walk on a
weighted directed graph on the vertex set V. Equivalently, it is an $n \times n$ probability matrix, for which the sum of each row is 1. A Markov chain is said to
be *ergodic* if the corresponding random walk converges to a unique stationary
distribution. That is, if there exists a vector π that is nonzero at each vertex, that satisfies $\pi = \pi M$, and such that for every vertex v in V, we have
$\lim_{t \to \infty} 1_v M^t = \pi$. The vector π is the *stationary distribution* of M. We remark
that a Markov chain is ergodic if and only if it is a random walk on a graph that
is strongly connected and aperiodic. Efficient numerical methods for computing
the stationary distribution of an ergodic Markov chain M are described in [16].

Let p be a probability distribution on the vertices of V, and let M be a Markov
chain. For each set $S \subseteq V$, we define the sum of p over S to be

$$p(S) = \sum_{u \in S} p(u),$$

For each edge (u, v), we define

$$p(u, v) = p(u)M(u, v).$$

This is the amount of probability that moves from u to v when a step of the
Markov chain is applied to the vector p. For each set A of directed edges, we
define

$$p(A) = \sum_{(u,v) \in A} p(u,v),$$

which is the total amount of probability moving over the set of directed edges. This notation is overloaded, but it is unambiguous if the type of input is known.

2.1   Conductance and Sweeps

We now assume that the Markov chain M is ergodic with a unique stationary distribution π, and define the generalizations to ergodic Markov chains of conductance, of the sweep procedure for finding cuts with small conductance (which is often used in spectral partitioning [4,15]), and of the potential function $p[x]$ (which was introduced by Lovàsz and Simonovits to bound the mixing rate of random walks). In the case of ergodic Markov chains, all of these are normalized by the stationary distribution π.

Given a set S of states, we define $\bar{\pi}(S) = min(\pi(S), 1 - \pi(S))$ to be the measure of the smaller side of the partition induced by S, and define the outgoing edge border $\partial(S)$ as follows,

$$\partial(S) = \{(u,v) \in E \mid u \in S \text{ and } v \in \bar{S}\}.$$

Definition 1. *Let M be an ergodic Markov chain, and let π be its unique stationary distribution. We define the M-conductance $\Phi_M(S)$ of a set of vertices S to be*

$$\Phi_M(S) = \frac{\pi(\partial(S))}{\bar{\pi}(S)}.$$

Definition 2. *Let M be an ergodic Markov chain with stationary distribution π, and let p be a probability distribution on the vertices. Let $v_1, \ldots, v_n$ be an ordering of the vertices such that*

$$\frac{p(v_i)}{\pi(v_i)} \geq \frac{p(v_{i+1})}{\pi(v_{i+1})}.$$

For each integer j in $\{1, \ldots, n\}$, we define $S_j^p = \{v_1, \ldots, v_j\}$ to be the set containing the top j vertices in this ordering. We define $\Phi_M(p)$ to be the smallest M-conductance among the sets $S_1^p, \ldots S_n^p$,

$$\Phi_M(p) = \min_{j \in [1,n]} \Phi_M(S_j^p).$$

The process of sorting the vertices according to this ordering and choosing the set of smallest M-conductance is called a sweep.

Definition 3. *Let M be an ergodic Markov chain with stationary distribution π, and let p be a probability distribution on the vertices. We define $p[x]$ to be the unique function from $[0,1]$ to $[0,1]$ such that*

$$p\left[\pi(S_j^p)\right] = p(S_j^p) \quad \text{for each } j \in [0, n],$$

and such that $p[x]$ is piecewise linear between these points.

Proposition 1. *We have the following facts about the function $p[x]$.*

1. *The function $p[x]$ is concave.*
2. *For any set S of vertices,*

$$p(S) \leq p\left[\pi(S)\right].$$

3. *For any set of directed edges A, we have*

$$p(A) \leq p\left[\pi(A)\right].$$

The facts in this proposition are proved in [11], and are not difficult to verify.

2.2   Global PageRank and Personalized PageRank

Definition 4. *Given a Markov chain M, the PageRank vector $\mathrm{pr}_M(\alpha, s)$, defined by Brin and Page [13], is the unique solution of the linear system*

$$\mathrm{pr}_M(\alpha, s) = \alpha s + (1 - \alpha)\mathrm{pr}_M(\alpha, s)M. \tag{1}$$

Here, α is a constant in $(0, 1]$ called the jump *probability, s is a probability distribution called the* starting *vector.*

We will use the following basic facts about PageRank.

Proposition 2. *For any Markov chain M, starting vector s, and jump probability $\alpha \in (0, 1]$, there is a unique vector $\mathrm{pr}_M(\alpha, s)$ satisfying*

$$\mathrm{pr}_M(\alpha, s) = \alpha s + (1 - \alpha)\mathrm{pr}_M(\alpha, s)M.$$

Proposition 3. *For any Markov chain M and any fixed value of α in $(0, 1]$, there is a linear transformation R_α such that $\mathrm{pr}_M(\alpha, s) = sR_\alpha$. Furthermore, R_α is given by the matrix*

$$R_\alpha = \alpha I + \alpha \sum_{t=1}^{\infty}(1 - \alpha)^t M^t. \tag{2}$$

We omit the proofs, which may be found elsewhere.

We let $\psi = \frac{1}{n}1_V$ be the uniform distribution. If a PageRank vector has ψ for its starting vector, we call it a *global PageRank vector*. If a PageRank vector has for its starting vector the indicator vector 1_v, with all probability on a single vertex v, we call it a *personalized PageRank vector*, and use the shorthand notation $\mathrm{pr}_M(\alpha, v) = \mathrm{pr}_M(\alpha, 1_v)$.

There are a plenitude of algorithms for computing global PageRank and personalized PageRank, so we will treat the computation of PageRank as a primitive operation. We assume we have the following two black-box algorithms,

- GlobalPR(M, α) computes the global PageRank vector $\mathrm{pr}_M(\alpha, \psi)$.
- LocalPR(M, α, v) computes the personalized PageRank vector $\mathrm{pr}_M(\alpha, v)$.

We make the distinction between these two black boxes because personalized PageRank can be computed more efficiently that global PageRank. One may use for LocalPR any of the algorithms described by Jeh and Widom [10], Berkhin [2], Sarlos [14], or Gleich [8], each of which can compute an approximation of the personalized PageRank vector $\mathrm{pr}_M(\alpha, v)$ by examining only a small fraction of the input graph near v, provided that M is a sparse matrix. The global PageRank can be computed efficiently in numerous ways, for example the Arnoldi method described in [9], but requires performing a computation over the entire graph. We will endeavor to use LocalPR instead of GlobalPR as much as possible.

3   Local Partitioning for Ergodic Markov Chains

We now state the main theorem of the paper, which shows that a sweep over a personalized PageRank vector in an ergodic Markov chain M can produce a set with small M-conductance. This is a natural generalization of the theorem proved for undirected graphs in [1].

Theorem 1. *Let M be an ergodic Markov chain with stationary distribution π. Let S be a set of vertices such that $\pi(S) \leq \frac{1}{2}$ and $\Phi_M(S) \leq \alpha/16$, for some constant α. If v is a vertex sampled from S according to the probability distribution $\pi(v)/\pi(S)$, then with probability at least $1/2$, we have $\Phi_M(\mathrm{pr}_M(\alpha, v)) = O(\sqrt{\alpha \log |S|})$.*

The proof of the theorem is given at the end of this section. Here is the outline of how we will proceed. Given a personalized PageRank vector $p = \mathrm{pr}_M(\alpha, s)$ in an ergodic Markov chain M, we place an upper bound on $p[x]$ that depends on α and $\Phi(p)$, and place a lower bound on $p[\pi(S)]$ that depends on the conductance of a certain set S near the starting vertex. These upper and lower bounds will be combined to show that $\Phi(p)$ is small. We establish the upper and lower bounds in the following lemmas.

Lemma 1. *Let M be an ergodic Markov chain with stationary distribution π, let $p = \mathrm{pr}_M(\alpha, v)$ be a personalized PageRank vector in M, and let $\phi = \Phi_M(p)$ be the smallest M-conductance found by the sweep over p. Then,*

$$p[x] \leq x + \alpha t + \left(1 - \tfrac{\phi^2}{72}\right)^t \sqrt{x/\pi(v)} \quad \text{for all } x \in [0, 1] \text{ and all } t \geq 0.$$

Lemma 2. *Let M be an ergodic Markov chain with stationary distribution π, let S be a set of vertices, and let v be a vertex sampled from S according to the probability distribution $\pi(v)/\pi(S)$. With probability at least $3/4$,*

$$\mathrm{pr}_M(\alpha, v)(S) \geq 1 - 4\frac{\Phi_M(S)}{\alpha}.$$

The proofs of these two lemmas are contained in the full version. We use them now to derive the main theorem.

Proof (**Proof of Theorem 1**)
Let $p = \text{pr}_M(\alpha, v)$ and let $\phi = \Phi(p)$. If v is sampled from S with probability $\pi(v)/\pi(S)$, Lemma 2 implies the following bound holds with probability at least $3/4$,

$$\text{pr}_M(\alpha, v)(S) \geq 1 - 4\frac{\Phi_M(S)}{\alpha} \geq 1 - 4\frac{\alpha/16}{\alpha} \geq 3/4. \tag{3}$$

We will now show that with probability at least $3/4$,

$$\frac{\pi(v)}{\pi(S)} \geq \frac{1}{4|S|}. \tag{4}$$

To see this, consider the set of vertices S' in S such that $\pi(v) \geq \frac{\pi(S)}{4|S|}$. Clearly $\pi(S \setminus S') < \pi(S)/4$, which shows that $\pi(S') > (3/4)\pi(S)$.

The probability that the two events described in (3) and (4) both occur is at least $1/2$. We will assume for the rest of the proof that both events hold.

Lemma 1 gives us the following upper bound on $\text{pr}_M(\alpha, v)(S)$.

$$\text{pr}_M(\alpha, v)(S) \leq \text{pr}_M(\alpha, v)[\pi(S)]$$

$$\leq (4/3)\pi(S) + \alpha T + \left(1 - \frac{\phi^2}{72}\right)^T \sqrt{\pi(S)/\pi(v)}$$

$$\leq (4/3)(1/2) + \alpha T + \left(1 - \frac{\phi^2}{72}\right)^T \sqrt{4|S|}.$$

If we let $T = (72/\phi^2)\ln 24\sqrt{4|S|}$, then

$$\text{pr}_M(\alpha, v)(S) \leq 2/3 + \alpha T + 1/24.$$

This contradicts our lower bound from (3) if $\alpha < 1/25T$, so we have shown that $\alpha \geq 1/25T$, which implies the following bound,

$$\phi \leq \sqrt{72 \cdot 25 \cdot \alpha \ln 24\sqrt{4|S|}} = O(\sqrt{\alpha \log |S|}).$$

4   Partitioning a Strongly Connected Graph

In the next two sections we describe two possible approaches to partitioning a directed graph. In this section, we describe the straightforward method that applies only when the directed graph is strongly connected.

If the graph is strongly connected, then we may apply Theorem 1 to the lazy random walk Markov chain $\mathcal{W}$, which is defined to be

$$\mathcal{W} = \mathcal{W}(A) = \frac{1}{2}(I + AD^{-1}).$$

Here, D is the diagonal matrix whose nonzero elements are the out-degrees of the vertices. The laziness of the walk ensures that $\mathcal{W}$ is ergodic whenever A is strongly connected, which allows us to apply our main theorem to $\mathcal{W}$.

To apply Theorem 1 to the lazy walk Markov chain $\mathcal{W}$, we must compute and perform a sweep over a personalized PageRank vector. When performing the sweep, we must know the stationary distribution of $\mathcal{W}$ to sort the vertices into the proper order. The stationary distribution needs to be computed only once, and afterwards we can find numerous cuts by computing a single personalized PageRank vector per cut. The necessary computation is summarized below.

Applying Theorem 1 to the lazy walk Markov chain of a strongly connected graph.

We are given as input a strongly connected directed graph with lazy walk matrix $\mathcal{W}$. The following procedure may be used to apply Theorem 1 with several different starting vertices and values of α. The offline preprocessing must be done once, after which the local computation may be performed as many times as desired.

Offline Preprocessing:

1. Compute the stationary distribution π of $\mathcal{W}$.

Local computation:

1. Pick a starting vertex v and a value of α.
2. Compute $p = \text{pr}_{\mathcal{W}}(\alpha, v)$, using `LocalPR`.
3. Sort the vertices in nonincreasing order of $p(x)/\pi(x)$.
4. Let S_j be the set of the top j vertices in this ranking.
5. Compute the $\mathcal{W}$-conductance of each set S_j^p, and output the set with the smallest $\mathcal{W}$-conductance.

---

5   Partitioning the PageRank Markov Chain

The majority of directed graphs that arise in practice are not strongly connected, so we cannot directly apply the results of the previous section to such a graph. In this section, we describe how Theorem 1 can be applied to the PageRank Markov chain of an arbitrary graph, which is always ergodic. We show that the notion of conductance associated with this Markov chain has a natural interpretation in terms of PageRank. We describe how to find a large number of sets with low conductance in the PageRank Markov chain by performing a small number (two) of global PageRank computations as a preprocessing step, followed by any desired number of local computations.

5.1   The PageRank Markov Chain

We now define the PageRank Markov chain $M_\beta = M_\beta(A)$ in terms of the adjacency matrix A of an arbitrary directed graph. To do so, we first modify the adjacency matrix by adding a self-loop to each vertex, to ensure that no vertex has out-degree zero. This ensures the random walk matrix $W = D^{-1}A$ is a Markov chain, where D is the diagonal matrix containing the modified out-degrees after the self-loops have been added.

Let $\psi = \frac{1}{n}\mathbf{1}_V$ be the uniform distribution, and let β be a constant in $[0, 1]$, which we will call the *global jump probability*. Recall that the global PageRank vector $\mathrm{pr}_W(\beta, \psi)$ is the unique solution of the linear system

$$\mathrm{pr}_W(\beta, \psi) = \beta\psi + (1 - \beta)\mathrm{pr}_W(\beta, \psi)W. \qquad (5)$$

The PageRank Markov chain M_β is defined to be

$$M_\beta = \beta K_\psi + (1 - \beta)W,$$

where $K_\psi = \mathbf{1}^T\psi$ is the dense rank-1 matrix obtained by taking the outer product of ψ with the all-ones vector. The global PageRank vector $\mathrm{pr}_W(\beta, \psi)$ is the stationary distribution of the PageRank Markov chain M_β. In other words, we have $\mathrm{pr}_W(\beta, \psi) = \mathrm{pr}_W(\beta, \psi)M_\beta$. The PageRank Markov chain M_β is ergodic for any value of $\beta \in (0, 1]$.

The notion of conductance associated with the PageRank Markov chain M_β has a natural interpretation in terms of the global PageRank vector $\mathrm{pr}_W(\beta, \psi)$. To describe this, we will use the shorthand notation $\mathrm{pr}_\beta = \mathrm{pr}_W(\beta, \psi)$ for the global PageRank, and $\Phi_\beta(S) = \Phi_{M_\beta}(S)$ for the M_β-conductance. Then, for any a set of vertices S, we have

$$\Phi_\beta(S) = \frac{\mathrm{pr}_\beta(\partial(S))}{\mathrm{pr}_\beta(S)}.$$

This is the probability that if we choose a vertex from S with probability proportional to its PageRank, and then take a single step in the PageRank Markov chain M_β, we end up at a vertex outside of S.

5.2 Computing Personalized PageRank in the PageRank Markov Chain

To apply our local partitioning theorem to M_β, we must compute a personalized PageRank vector in the Markov chain M_β. The personalized PageRank vector $\mathrm{pr}_{M_\beta}(\alpha, s)$ is the unique solution of the linear system

$$\mathrm{pr}_{M_\beta}(\alpha, s) = \alpha s + (1 - \alpha)\mathrm{pr}_{M_\beta}(\alpha, s)M_\beta.$$

Although this is a personalized PageRank vector, the Markov chain M_β is dense because of its global random jump, so it is not possible to compute $\mathrm{pr}_{M_\beta}(\alpha, s)$ efficiently using $\texttt{LocalPR}(M_\beta, \alpha, s)$. We will show that $\mathrm{pr}_{M_\beta}(\alpha, s)$ can be computed efficiently in another way, by taking a linear combination of a personalized PageRank vector and a global PageRank vector in the random walk Markov chain W.

We now present two interpretations of the PageRank vector $\mathrm{pr}_{M_\beta}(\alpha, s)$. By definition, $\mathrm{pr}_{M_\beta}(\alpha, s)$ is a personalized PageRank vector in the Markov chain M_β. It can also be viewed as a PageRank vector in the random walk Markov chain W. When viewed as a PageRank vector in W, its starting vector is a linear combination of the uniform distribution ψ and the starting vector s, and its jump probability is $\gamma = \alpha + \beta - \alpha\beta$.

$$\begin{aligned}
\mathrm{pr}_{M_\beta}(\alpha, s) &= \alpha s + (1-\alpha)\mathrm{pr}_\beta(\alpha, s)M_\beta \\
&= \alpha s + (1-\alpha)\beta\psi + (1-\alpha)(1-\beta)\mathrm{pr}_\beta(\alpha, s)W \\
&= \gamma\left(\frac{\alpha}{\gamma}s + \frac{(1-\alpha)\beta}{\gamma}\psi\right) + (1-\gamma)\mathrm{pr}_\beta(\alpha, s)W \\
&= \mathrm{pr}_W(\gamma, s').
\end{aligned}$$

Here $\gamma = \alpha + \beta - \alpha\beta$, and $s' = \frac{\alpha}{\gamma}s + \frac{(1-\alpha)\beta}{\gamma}\psi$. Using the fact that a PageRank vector is a linear function of its starting vector, we can write

$$\begin{aligned}
\mathrm{pr}_{M_\beta}(\alpha, s) &= \mathrm{pr}_W(\gamma, \frac{\alpha}{\gamma}s + \frac{(1-\alpha)\beta}{\psi}) \\
&= \frac{\alpha}{\gamma}\mathrm{pr}_W(\gamma, s) + \frac{(1-\alpha)\beta}{\gamma}\mathrm{pr}_W(\gamma, \psi).
\end{aligned}$$

In summary, we have taken a personalized PageRank vector $\mathrm{pr}_{M_\beta}(\alpha, s)$ from the PageRank Markov chain M_β, and written it as a linear combination of two PageRank vectors from the walk Markov chain W. One of these is a personalized PageRank vector in W with starting vector s, and the other is a global PageRank vector in W with starting distribution ψ.

5.3 Local Partitioning in the PageRank Markov Chain

By applying our main theorem to the PageRank Markov chain, we obtain the following corollary, which shows a sweep over the PageRank vector $\mathrm{pr}_{M_\beta}(\alpha, v)$ produces a set with small M_β-conductance.

Corollary 1. *Let S be a set of vertices such that $\mathrm{pr}_\beta(S) \leq \frac{1}{2}$ and $\Phi_\beta(S) \leq \alpha/16$, for some constants α and β. If a vertex v is sampled from S according to the probability distribution $\mathrm{pr}_\beta(v)/\mathrm{pr}_\beta(S)$, then with probability at least $1/2$ we have $\Phi_\beta(\mathrm{pr}_{M_\beta}(\alpha, v)) = O(\sqrt{\alpha \log |S|})$.*

Proof. The corollary is immediate, by applying Theorem 1 to the ergodic Markov chain M_β.

To carry out the computation required by the corollary, we need to compute the stationary distribution of M_β, which is just the global PageRank vector $\mathrm{pr}_W(\beta, \psi)$. For each cut we want to find, we also need to compute a personalized PageRank vector $\mathrm{pr}_{M_\beta}(\alpha, v)$ in the Markov chain M_β. This can be done by computing $\mathrm{pr}_W(\gamma, v)$ and $\mathrm{pr}_W(\gamma, \psi)$, and then taking a linear combination of these two PageRank vectors, as described in the previous section. If we fix the values of α and β, we can compute the two global PageRank vectors $\mathrm{pr}_W(\beta, \psi)$ and $\mathrm{pr}_W(\gamma, \psi)$ ahead of time, and then compute a large number of personalized PageRank vectors $\mathrm{pr}_W(\gamma, v)$ using LocalPR. This procedure is summarized below.

Applying Corollary 1 to the PageRank Markov chain.
We are given as input the adjacency matrix A of a directed graph (not necessarily strongly connected), the global jump probability β, and the local jump probability α. The following procedure may be used to apply Theorem 1 at several different starting vertices with these fixed values of α and β. The offline preprocessing must be done once, after which the local computation may be performed as many times as desired.

Offline Preprocessing:
We must compute two global PageRank vectors.

1. Let $\gamma = \alpha + \beta - \alpha\beta$.
2. Let $W = W(A)$ be the random walk matrix of A.
3. Compute the two global PageRank vectors $\mathrm{pr}_\beta = \mathrm{pr}_W(\beta, \psi)$ and $\mathrm{pr}_\gamma = \mathrm{pr}_W(\gamma, \psi)$ using the algorithm GlobalPR.

Local Computation:

1. Pick a starting vertex v.
2. Compute $\mathrm{pr}_W(\gamma, v)$, using LocalPR.
3. Obtain $p = \mathrm{pr}_{M_\beta}(\alpha, v)$ by taking a linear combination of $\mathrm{pr}_W(\gamma, v)$ and $\mathrm{pr}_W(\gamma, \psi)$,

$$
p = \mathrm{pr}_{M_\beta}(\alpha, v) = \frac{\alpha}{\gamma}\mathrm{pr}_W(\gamma, v) + \frac{(1-\alpha)\beta}{\gamma}\mathrm{pr}_W(\gamma, \psi).
$$

4. Rank the vertices in nonincreasing order of $p(x)/\mathrm{pr}_\beta(x)$.
5. Let S_j be the set of the top j vertices in this ranking.
6. Compute the β-conductances $\Phi_\beta(S_j)$ for each set S_j, and output the set with the smallest β-conductance.

---

6   Concluding Remarks

6.1   When Is Partitioning the PageRank Markov Effective?

Corollary 1 can be applied to partition the PageRank Markov chain of an arbitrary directed graph, and to an arbitrary starting vertex. Because it may be applied to any graph (even the empty graph), the approximation guarantee it provides may become vacuous for some graphs and starting vertices. In this section we will describe this concern in more detail, and give a positive result that describes when the approximation guarantee it provides is strong rather than vacuous. We caution that this section contains high-level discussion rather than rigorous proofs.

As we increase β, we increase the probability of the global jump, which ensures that the β-conductance of every set in the graph is at least roughly β. If we partition the PageRank Markov chain of a graph with no edges, every subset of vertices will have conductance roughly β, so the approximation guarantee of Corollary 1 will be vacuous (which is what we should expect when partitioning a graph with no edges). On the other hand, if we partition the PageRank Markov chain of an undirected graph, using a very small value of β, the best partitions of the graph will have β-conductance larger than β, so the approximation guarantee of Corollary 1 will give a meaningful result.

Loosely speaking, we claim that partitioning the PageRank Markov chain M_β gives interesting results exactly when there are interesting partitions of the graph that have β-conductance larger than β. To provide evidence for this claim, we separate the β-conductance $\Phi_\beta(S)$ into two parts, the contribution $\Psi_\beta(S)$ from real graph edges in W, and the contribution from the random jump. We define

$$\Psi_\beta(S) = \frac{\sum_{(u,v)\in S\times\bar{S}} \mathrm{pr}_\beta(u)W(u,v)}{\mathrm{pr}_\beta(S)}.$$

Then, $\Phi_\beta(S)$ and $\Psi_\beta(S)$ are related by the following equation.

$$\Phi_\beta(S) = (1-\beta)\Psi_\beta(S) + \beta\frac{|\bar{S}|}{n}.$$

It is not hard to see that if a set S has β-conductance significantly larger than β, our algorithm finds a set S' for which $\Psi_\beta(S')$ is nearly as small as $\Psi_\beta(S)$. In particular, if S is a set of vertices for which $\Psi_\beta(S) = \Omega(\Phi_\beta(S))$, and S' is a set of vertices for which $\Phi_\beta(S') = O(\sqrt{\Phi_\beta(S)\log n})$, which is the conductance guaranteed by Corollary 1, then we have

$$\Psi_\beta(S') = O(\sqrt{\Psi_\beta(S)\log n}).$$

6.2 Cuts from Approximate PageRank Vectors

For the case of undirected graphs, it has been proved that a cut with small conductance can be found efficiently by sweeping over an *approximate* personalized PageRank vector. This was proved in [1], and requires a careful error analysis. We remark that a similar error analysis may be carried out for the directed case, although we have not described such an analysis in this paper.

Acknowledgements

We thank Zoltán Gyöngyi for his expert advice during the formative stages of this research.

References

1. Andersen, R., Chung, F., Lang, K.: Local graph partitioning using PageRank vectors. In: FOCS 2006: Proceedings of the 47th Annual IEEE Symposium on Foundations of Computer Science, pp. 475–486. IEEE Computer Society, Washington, DC (2006)
2. Berkhin, P.: Bookmark-coloring algorithm for personalized PageRank computing. Internet Math. 3(1), 41–62 (2006)
3. Charikar, M., Makarychev, K., Makarychev, Y.: Directed metrics and directed graph partitioning problems. In: SODA 2006: Proceedings of the seventeenth annual ACM-SIAM symposium on Discrete algorithm, pp. 51–60. ACM Press, New York (2006)
4. Chung, F.: Spectral graph theory. In: CBMS Regional Conference Series in Mathematics, vol. 92, American Mathematical Society, Providence (1997)
5. Chung, F.: Laplacians and Cheeger inequalities for directed graphs. Annals of Combinatorics 9, 1–19 (2005)
6. Chuzhoy, J., Khanna, S.: Hardness of cut problems in directed graphs. In: STOC 2006: Proceedings of the thirty-eighth annual ACM symposium on Theory of computing, pp. 527–536. ACM Press, New York (2006)
7. Fill, J.A.: Eigenvalue bounds on convergence to stationarity for nonreversible Markov chains, with an application to the exclusion process. Ann. Appl. Probab. 1(1), 62–87 (1991)
8. Gleich, D., Polito, M.: Approximating personalized PageRank with minimal use of webgraph data. Internet Mathematics (to appear)
9. Golub, G., Greif, C.: Arnoldi-type algorithms for computing stationary distribution vectors, with application to PageRank. 10543 BIT Numerical Mathematics 46(4) (2006)
10. Jeh, G., Widom, J.: Scaling personalized web search. In: WWW 2003. Proceedings of the 12th World Wide Web Conference, pp. 271–279 (2003)
11. Lovász, L., Simonovits, M.: The mixing rate of markov chains, an isoperimetric inequality, and computing the volume. In: FOCS, pp. 346–354 (1990)
12. Mihail, M.: Conductance and convergence of markov chains—a combinatorial treatment of expanders. In: Proc. of 30th FOCS, pp. 526–531 (1989)
13. Page, L., Brin, S., Motwani, R., Winograd, T.: The PageRank citation ranking: Bringing order to the web. Technical report, Stanford Digital Library Technologies Project (1998)
14. Sarlós, T., Benczúr, A.A., Csalogány, K., Fogaras, D.: To randomize or not to randomize: space optimal summaries for hyperlink analysis. In: WWW, pp. 297–306 (2006)
15. Spielman, D.A., Teng, S.-H.: Spectral partitioning works: Planar graphs and finite element meshes. In: IEEE Symposium on Foundations of Computer Science, pp. 96–105 (1996)
16. Stewart, W.: Introduction to the Numerical Solution of Markov Chains. Princeton Univ. Press (1994)

Stochastic Kronecker Graphs

Mohammad Mahdian[1] and Ying Xu[2,*]

[1] Yahoo! Research
mahdian@yahoo-inc.com
[2] Stanford University
xuying@cs.stanford.edu

Abstract. A random graph model based on Kronecker products of probability matrices has been recently proposed as a generative model for large-scale real-world networks such as the web. This model simultaneously captures several well-known properties of real-world networks; in particular, it gives rise to a heavy-tailed degree distribution, has a low diameter, and obeys the densification power law. Most properties of Kronecker products of graphs (such as connectivity and diameter) are only rigorously analyzed in the deterministic case. In this paper, we study the basic properties of stochastic Kronecker products based on an initiator matrix of size two (which is the case that is shown to provide the best fit to many real-world networks). We will show a phase transition for the emergence of the giant component and another phase transition for connectivity, and prove that such graphs have constant diameters beyond the connectivity threshold, but are not searchable using a decentralized algorithm.

1 Introduction

A generative model based on Kronecker matrix multiplication was recently proposed by Leskovec et al. [10] as a model that captures many properties of real-world networks. In particular, they observe that this model exhibits a heavy-tailed degree distribution, and has an average degree that grows as a power law with the size of the graph, leading to a diameter that stays bounded by a constant (the so-called *densification power law* [12]). Furthermore, Leskovec and Faloutsos [11] fit the stochastic model to some real world graphs, such as Internet Autonomous Systems graph and Epinion trust graphs, and find that Kronecker graphs with appropriate 2×2 initiator matrices mimic very well many properties of the target graphs.

Most properties of the Kronecker model (such as connectivity and diameter) are only rigorously analyzed in the deterministic case (i.e., when the initiator matrix is a binary matrix, generating a single graph, as opposed to a distribution over graphs), and empirically shown in the general stochastic case [10]. In this paper we analyze some basic graph properties of stochastic Kronecker graphs with an initiator matrix of size 2. This is the case that is shown by Leskovec

---

* Work performed in part while visiting Yahoo! Research.

A. Bonato and F.R.K. Chung (Eds.): WAW 2007, LNCS 4863, pp. 179–186, 2007.
© Springer-Verlag Berlin Heidelberg 2007

and Faloutsos [11] to provide the best fit to many real-world networks. We give necessary and sufficient conditions for Kronecker graphs to be connected or to have giant components of size $\Theta(n)$ with high probability[1]. Our analysis of the connectivity of Kronecker graphs is based on a general lemma (Theorem 1) that might be of independent interest. We prove that under the parameters that the graph is connected with high probability, it also has a constant diameter with high probability. This unusual property is consistent with the observation of Leskovec et al. [12] that in many real-world graphs the effective diameters do not increase, or even shrink, as the sizes of the graphs increase, which is violated by many other random graph models with increasing diameters. Finally we show that Kronecker graphs do not admit short (poly-logarithmic) routing paths by decentralized routing algorithms based on only local information.

1.1   The Model and Overview of Results

In this paper we analyze stochastic Kronecker graphs with an initiator matrix of size 2, as defined below:

Definition 1. *A* (stochastic) Kronecker graph *is defined by*

(i) an integer k, and
(ii) a symmetric 2×2 *matrix* θ*:* $\theta[1,1] = \alpha, \theta[1,0] = \theta[0,1] = \beta, \theta[0,0] = \gamma$,

where $0 \leq \gamma \leq \beta \leq \alpha \leq 1$. *We call* θ *the base* matrix *or the* initiator *matrix.*
    The graph has $n = 2^k$ *vertices, each vertex labeled by a unique bit vector of length k; given two vertices u with label* $u_1 u_2 \ldots u_k$ *and v with label* $v_1 v_2 \ldots v_k$, *the probability of edge* (u,v) *existing, denoted by* $P[u,v]$, *is* $\prod_i \theta[u_i, v_i]$, *independent on the presence of other edges.*

In particular, when $\alpha = \beta = \gamma$, the Kronecker graph becomes the well studied random graph $G(n, p)$ with $p = \alpha^k$. Leskovec and Faloutsos [11] showed that the Kronecker graph model with 2×2 initiator matrices satisfying the above conditions is already very powerful in simulating real world graphs. In fact, their experiment shows that the matrix [.98, .58; .58, .06] is a good fit for the Internet AS graph. When the base matrix does not satisfy the condition stated in the above definition (i.e., when $\alpha \geq \gamma \geq \beta$ or $\beta \geq \alpha \geq \gamma$), Kronecker graphs appear to have different structural properties, and require different analytic techniques. We can prove some of our results in these regimes as well; however, due to lack of space and since this setting of parameters is less appealing as a generative model for the web, we omit the results in this paper; more detail can be found in our technical report [13].
    We analyze basic graph properties of the stochastic Kronecker graph model. In particular, we prove that the necessary and sufficient condition for Kronecker graphs to be connected with high probability is $\beta + \gamma > 1$ or $\alpha = \beta = 1, \gamma = 0$ (Section 2.2); the necessary and sufficient condition for Kronecker graphs to have a giant component of size $\Theta(n)$ with high probability is $(\alpha + \beta)(\beta + \gamma) > 1$, or

---

[1] Throughout the paper by "with high probability" we mean with probability $1 - o(1)$.

$(\alpha + \beta)(\beta + \gamma) = 1$ and $\alpha + \beta > \beta + \gamma$ (Section 2.3); if $\beta + \gamma > 1$, the diameters of Kronecker graphs are constant with high probability (Section 3); and that no decentralized search algorithm can find a path of length $o(n^{(1-\alpha)\log_2 e})$ between a given pair of vertices in Kronecker graphs, unless the graph is deterministic (Section 4).

Besides Kronecker graphs, we also define a general family of random graphs $G(n, P)$, which generalizes all random graph models where edges are independent, including Kronecker graphs and $G(n, p)$.

Definition 2. *A random graph $G(n, P)$, where n is an integer and P is an $n \times n$ matrix with elements in $[0, 1]$, has n vertices and includes each edge (i, j) independently with probability $P[i, j]$.*

Throughout this paper we consider undirected $G(n, P)$: P is symmetric and edges are undirected. We prove two useful theorems about connectivity and searchability in this model, which may be of independent interest; namely, we show that if the min-cut size of the weighted graph defined by P is at least $c \ln n$ (c is a sufficiently large constant), then with high probability $G(n, P)$ is connected (Section 2.1), and prove a monotonicity property for searchability in this model (Section 4).

2 Connectivity and Giant Components

We first state a sufficient condition for connectivity of general random graphs $G(n, P)$ (Section 2.1), then use this condition to analyze connectivity and giant components of Kronecker graphs (Section 2.2, 2.3).

2.1 Connectivity of $G(n, P)$

We give a sufficient condition of the matrix P for $G(n, P)$ graphs to be connected. Let V be the set of all vertices. For any $S, S' \subseteq V$, define $P(i, S) = \sum_{j \in S} P[i, j]$; $P(S, S') = \sum_{i \in S, j \in S'} P[i, j]$.

Theorem 1. *If the min-cut size of the weighted graph defined by P is $c \ln n$ (c is a sufficiently large constant), i.e. $\forall S \subset V, P(S, V \setminus S) \geq c \ln n$, then with high probability $G(n, P)$ is connected.*

Proof. A k-*minimal cut* is a cut whose size is at most k times the min-cut size. We use the following result about the number of k-minimal cuts due to Karger and Stein [5]: In any weighted graph, the number of k-minimal cuts is at most $O((2n)^{2k})$.

Consider the weighted graph defined by P. Denote its min-cut size by t. We say a cut is a k-cut if its size is between kt and $(k + 1)t$. By the above result there are at most $O((2n)^{2k+2})$ k-cuts. Now consider a fixed k-cut in a random realization of $G(n, P)$: the expected size of the cut is at least kt, so by Chernoff bound the probability that the cut has size 0 in the realization is at most $e^{-kt/2}$.

Taking the union bound over all k-cuts, for all $k = 1, 2, \ldots, n^2$, the probability that at least one cut has size 0 is bounded by

$$\sum_{k=1,\ldots,n^2} e^{-kt/2} O((2n)^{2k+2})$$

For $t = c \ln n$ where c is a sufficiently large constant, this probability is $o(1)$. Therefore with high probability $G(n, P)$ is connected.

Note that $G(n, p)$ is known to be disconnected with high probability when $p \leq (1 - \epsilon) \ln n / n$, i.e., when the min-cut size is $(1 - \epsilon) \ln n$. Therefore the condition in the above theorem is tight up to a constant factor. Also, extrapolating from $G(n, p)$, one might hope to prove a result similar to the above for the emergence of the giant component; namely, if the size of the min-cut in the weighted graph defined by P is at least a constant, $G(n, P)$ has a giant component. However, this result is false, as can be seen from this example: n vertices are arranged on a cycle, and P assigns a probability of 0.5 to all pairs that are within distance c (a constant) on the cycle, and 0 to all other pairs. It is not hard to prove that with high probability $G(n, P)$ does not contain any connected component of size larger than $O(\log n)$.

2.2   Connectivity of Kronecker Graphs

We define the *weight* of a vertex to be the number of 1's in its label; denote the vertex with weight 0 by **0**, and the vertex with weight k by **1**. We say a vertex u is *dominated* by vertex u', denoted by $u \leq u'$, if for any bit i, $u_i \leq u'_i$. Recall that $P[u, v]$ is as defined in Definition 1.

The following lemmas state some simple facts about Kronecker graphs. Lemma 1 is trivially true given the condition $\alpha \geq \beta \geq \gamma$. The proof of Lemma 2 is omit due to the lack of space and can be found in our technical report [13].

Lemma 1. *For any vertex u, $\forall v, P[u, v] \geq P[\mathbf{0}, v]; \forall S, P(u, S) \geq P(\mathbf{0}, S)$. Generally, for any vertices $u \leq u'$, $\forall v, P[u, v] \leq P[u', v]; \forall S, P(u, S) \leq P(u', S)$.*

Lemma 2. *The expected degree of a vertex u with weight l is $(\alpha + \beta)^l (\beta + \gamma)^{k-l}$.*

Theorem 2. *The necessary and sufficient condition for Kronecker graphs to be connected with high probability (for large k) is $\beta + \gamma > 1$ or $\alpha = \beta = 1, \gamma = 0$.*

Proof. We first show that this is a necessary condition for connectivity.

Case 1. If $\beta + \gamma < 1$, the expected degree of vertex **0** is $(\beta + \gamma)^k = o(1)$, with high probability vertex **0** is isolated and the graph is thus disconnected.

Case 2. If $\beta + \gamma = 1$ but $\beta < 1$, we again prove that with constant probability vertex **0** is isolated:

$$Pr[\mathbf{0} \text{ has no edge}] = \prod_v (1 - P[\mathbf{0}, v]) = \prod_{w=0}^{k} (1 - \beta^w \gamma^{k-w})^{\binom{k}{w}} \geq \prod_{w=0}^{k} e^{-2\binom{k}{w}\beta^w \gamma^{k-w}}$$

$$= e^{-2 \sum_{w=0}^{k} \binom{k}{w} \beta^w \gamma^{k-w}} = e^{-2(\beta+\gamma)^k} = e^{-2}$$

Now we prove it is also a sufficient condition. When $\alpha = \beta = 1, \gamma = 0$, the graph embeds a deterministic star centered at vertex **1**, and is hence connected. To prove $\beta + \gamma > 1$ implies connectivity, we only need to show the min-cut has size at least $c \ln n$ and apply Theorem 1. The expected degree of vertex **0** excluding self-loop is $(\beta + \gamma)^k - \gamma^k > 2ck = 2c \ln n$ given that β and γ are constants independent on k satisfying $\beta + \gamma > 1$, therefore the cut $(\{\mathbf{0}\}, V \setminus \{\mathbf{0}\})$ has size at least $2c \ln n$. Remove **0** and consider any cut $(S, V \setminus S)$ of the remaining graph, at least one side of the cut gets at least half of the expected degree of vertex **0**; without loss of generality assume it is S i.e. $P(\mathbf{0}, S) > c \ln n$. Take any node u in $V \setminus S$, by Lemma 1, $P(u, S) \geq P(\mathbf{0}, S) > c \ln n$. Therefore the cut size $P(S, V \setminus S) \geq P(u, S) > c \ln n$.

2.3   Giant Components

Lemma 3. *Let H denote the set of vertices with weight at least $k/2$, then for any vertex u, $P(u, H) \geq P(u, V)/4$.*

Proof. Given u, let l be the weight of u. For a vertex v let $i(v)$ be the number of bits where $u_b = v_b = 1$, and let $j(v)$ be the number of bits where $u_b = 0, v_b = 1$. we partition the vertices in $V \setminus H$ into 3 subsets: $S_1 = \{v : i(v) \geq l/2, j(v) < (k - l)/2\}, S_2 = \{v : i(v) < l/2, j(v) \geq (k - l)/2\}, S_3 = \{v : i(v) < l/2, j(v) < (k - l)/2\}$.

First consider S_1. For a vertex $v \in S_1$, we flip the bits of v where the corresponding bits of u is 0 to get v'. Then $i(v') = i(v)$ and $j(v') \geq (k - l)/2 > j(v)$. It is easy to check that $P[u, v'] \geq P[u, v]$, $v' \in H$, and different $v \in S_1$ maps to different v'. Therefore $P(u, H) \geq P(u, S_1)$.

Similarly we can prove $P(u, H) \geq P(u, S_2)$ by flipping the bits corresponding to 1s in u, and $P(u, H) \geq P(u, S_3)$ by flipping all the bits. Adding up the three subsets, we get $P(u, V \setminus H) \leq 3P(u, H)$. Thus, $P(u, H) \geq P(u, V)/4$.

Theorem 3. *The necessary and sufficient condition for Kronecker graphs to have a giant component of size $\Theta(n)$ with high probability is $(\alpha + \beta)(\beta + \gamma) > 1$, or $(\alpha + \beta)(\beta + \gamma) = 1$ and $\alpha + \beta > \beta + \gamma$.*

Proof. When $(\alpha + \beta)(\beta + \gamma) < 1$, we prove that the expected number of non-isolated nodes are $o(n)$. Let $(\alpha + \beta)(\beta + \gamma) = 1 - \epsilon$. Consider vertices with weight at least $k/2 + k^{2/3}$, by Chernoff bound the fraction of such vertices is at most $\exp(-ck^{4/3}/k) = \exp(-ck^{1/3}) = o(1)$, therefore the number of non-isolated vertices in this category is $o(n)$; on the other hand, for a vertex with weight less than $k/2 + k^{2/3}$, by Lemma 2 its expected degree is at most

$$(\alpha + \beta)^{k/2+k^{2/3}}(\beta + \gamma)^{k/2-k^{2/3}} = (1 - \epsilon)^{k/2}\left(\frac{\alpha + \beta}{\beta + \gamma}\right)^{k^{2/3}} = n^{-\epsilon'}c^{o(\log n)} = o(1)$$

Therefore overall there are $o(n)$ non-isolated vertices.

When $\alpha + \beta = \beta + \gamma = 1$, i.e. $\alpha = \beta = \gamma = 1/2$, the Kronecker graph is equivalent to $G(n, 1/n)$, which has no giant component of size $\Theta(n)$ [4].

When $(\alpha + \beta)(\beta + \gamma) > 1$, we prove that the subgraph induced by $H = \{v : weight(v) \geq k/2\}$ is connected with high probability, hence forms a giant connected component of size at least $n/2$. Again we prove that the min-cut size of H is $c \ln n$ and apply Theorem 1. For any vertex u in H, its expected degree is at least $((\alpha + \beta)(\beta + \gamma))^{k/2} = \omega(\ln n)$; by Lemma 3 $P(u, H) \geq P(u, V)/4 > 2c \ln n$. Now given any cut $(S, H \setminus S)$ of H, we prove $P(S, H \setminus S) > c \ln n$. Without loss of generality assume vertex $\mathbf{1}$ is in S. For any vertex $u \in H$, either $P(u, S)$ or $P(u, H \setminus S)$ is at least $c \ln n$. If $\exists u$ such that $P(u, H \setminus S) > c \ln n$, then since $u \leq \mathbf{1}$, by Lemma 1 $P(S, H \setminus S) \geq P(\mathbf{1}, H \setminus S) \geq P(u, H \setminus S) > c \ln n$; otherwise $\forall u \in H, P(u, S) > c \ln n$, since at least one of the vertex is in $H \setminus S$, $P(S, H \setminus S) > c \ln n$.

Finally, when $(\alpha + \beta)(\beta + \gamma) = 1$ and $\alpha + \beta > \beta + \gamma$, let $H_1 = \{v : weight(v) \geq k/2 + k^{1/6}\}$, and we will prove that the subgraph induced by H_1 is connected with high probability by proving its min-cut size is at least $c \ln n$ (Claim 1), and that $|H_1| = \Theta(n)$ (Claim 2), therefore with high probability H_1 forms a giant connected component of size $\Theta(n)$. The proofs of these claims is omit due to the lack of space and can be found in our technical report [13].

3   Diameter

We analyze the diameter of a Kronecker graph under the condition that the graph is connected with high probability. When $\alpha = \beta = 1, \gamma = 0$, every vertex links to $\mathbf{1}$ so the graph has diameter 2; below we analyze the case where $\beta + \gamma > 1$. We will use the following result about the diameter of $G(n, p)$, which has been extensively studied in for example [6,2,3].

Theorem 4. [6,2] If $(pn)^{d-1}/n \to 0$ and $(pn)^d/n \to \infty$ for a fixed integer d, then $G(n, p)$ has diameter d with probability approaching 1 as n goes to infinity.

Theorem 5. If $\beta + \gamma > 1$, the diameters of Kronecker graphs are constant with high probability.

Proof. Let S be the subset of vertices with weight at least $\frac{\beta}{\beta + \gamma} k$. We will prove that the subgraph induced by S has a constant diameter, and any other vertex directly connects to S with high probability.

Claim 3. With high probability, any vertex u has a neighbor in S.

Proof (Proof of Claim 3.). We compute the expected degree of u to S:

$$P(u, S) \geq \sum_{j \geq \frac{\beta}{\beta+\gamma}k} \binom{k}{j} \beta^j \gamma^{k-j} = (\beta + \gamma)^k \sum_{j \geq \frac{\beta}{\beta+\gamma}k} \binom{k}{j} \left(\frac{\beta}{\beta+\gamma}\right)^j \left(\frac{\gamma}{\beta+\gamma}\right)^{k-j}$$

The summation is exactly the probability of getting at least $\frac{\beta}{\beta+\gamma} k$ HEADs in k coin flips where the probability of getting HEAD in one trial is $\frac{\beta}{\beta+\gamma}$, so this probability is at least a constant. Therefore $P(u, S) \geq (\beta + \gamma)^k/2 > c \ln n$ for any u; by Chernoff bound any u has a neighbor in S with high probability.

Claim 4. $|S| \cdot \min_{u,v \in S} P[u,v] \geq (\beta + \gamma)^k$.

Proof (Proof of Claim 4.). We have

$$\min_{u,v \in S} P[u,v] \geq \beta^{\frac{\beta}{\beta+\gamma}k} \gamma^{\frac{\gamma}{\beta+\gamma}k}$$

and

$$|S| \geq \binom{k}{\frac{\beta}{\beta+\gamma}k} \approx \frac{(\frac{k}{e})^k}{(\frac{\beta k}{(\beta+\gamma)e})^{\frac{\beta}{\beta+\gamma}k}(\frac{\gamma k}{(\beta+\gamma)e})^{\frac{\gamma}{\beta+\gamma}k}} = \frac{(\beta+\gamma)^k}{\beta^{\frac{\beta}{\beta+\gamma}k}\gamma^{\frac{\gamma}{\beta+\gamma}k}}$$

Therefore $|S| \cdot \min_{u,v \in S} P[u,v] \geq (\beta + \gamma)^k$.

Given Claim 4, it follows easily that the diameter of the subgraph induced by S is constant: let $\beta + \gamma = 1 + \epsilon$ where ϵ is a constant, the diameter of $G(|S|, (\beta + \gamma)^k/|S|)$ is at most $d = 1/\epsilon$ by Theorem 4; since by increasing the edge probabilities of $G(n, P)$ the diameter cannot increase, the diameter of the subgraph of the Kronecker graph induced by S is no larger than that of $G(|S|, (\beta+\gamma)^k/|S|)$. Therefore, by Claim 3, for every two vertices u and v in the Kronecker graph, there is a path of length at most $2 + 1/\epsilon$ between them.

4   Searchability

In Section 3 we showed that the diameter of a Kronecker graph is constant with high probability, given that the graph is connected. However it is yet a question whether a short path can be found by a decentralized algorithm where each individual only has access to local information. We use a similar definition as used by Kleinberg [7,8,9].

Definition 3. *In a decentralized routing algorithm for $G(n, P)$, the message is passed sequentially from a current message holder to one of its neighbors until reach the destination t, using only local information. In particular, the message holder u at a given step has knowledge of:*

(i) the probability matrix P;
(ii) the label of destination t;
(iii) edges incident to all visited vertices.

A $G(n, P)$ graph is d-searchable *if there exists a decentralized routing algorithm such that for any destination t, source s, with high probability the algorithm can find an s-t path no longer than d.*

We first give a monotonicity result on general random graphs $G(n, P)$, then use it to prove Kronecker graphs with $\alpha < 1$ is not poly-logarithmic searchable. It is possible to directly prove our result on Kronecker graphs, but we believe the monotonicity theorem might be of independent interests. More results on searchability in $G(n, P)$ using deterministic memoryless algorithms can be found in [1]. The proof of Theorem 6 can be found in our technical report [13].

Theorem 6. *If $G(n, P)$ is d-searchable, and $P \leq P'$ ($\forall i, j, P[i,j] \leq P'[i,j]$), then $G(n, P')$ is d-searchable.*

Theorem 7. *Kronecker graphs are not $n^{(1-\alpha)\log_2 e}$-searchable.*

Proof. Let P be the probability matrix of the Kronecker graph, and P' be the matrix where each element is $p = n^{-(1-\alpha)\log_2 e}$. We have $P \leq P'$ because $max_{i,j} P[i,j] \leq \alpha^k \leq e^{-(1-\alpha)k} = n^{-(1-\alpha)\log_2 e} = p$. If the Kronecker graph is $n^{(1-\alpha)\log_2 e}$-searchable, then by Theorem 6 $G(n, p)$ where $p = n^{-(1-\alpha)\log_2 e}$ is also $n^{(1-\alpha)\log_2 e}$-searchable. However, $G(n, p)$ is not $\frac{1}{p}$-searchable. This is because given any decentralized algorithm, whenever we first visit a vertex u, independent on the routing history, the probability that u has a direct link to t is no more than p, hence the routing path is longer than the geometry distribution with parameter p, i.e. with constant probability the algorithm cannot reach t in $1/p$ steps.

References

1. Arcaute, E., Chen, N., Kumar, R., Liben-Nowell, D., Mahdian, M., Nazerzadeh, H., Xu, Y.: Searchability in random graphs. In: Proceedings of the 5th Workshop on Algorithms and Models for the Web-Graph (2007)
2. Bollobás, B.: The diameter of random graphs. IEEE Trans. Inform. Theory 36(2), 285–288 (1990)
3. Chung, F., Lu, L.: The diameter of random sparse graphs. Advances in Applied Math. 26, 257–279 (2001)
4. Erdös, P., Rényi, A.: On random graphs I. Publications Mathematics, Debrecen 6, 290–297 (1959)
5. Karger, D., Stein, C.: A new approach to the minimum cut problem. Journal of the ACM 43(4) (1996)
6. Klee, V., Larmann, D.: Diameters of random graphs. Canad. J. Math. 33, 618–640 (1981)
7. Kleinberg, J.: The small-world phenomenon: An algorithmic perspective. In: Proc. 32nd ACM Symposium on Theory of Computing (2000)
8. Kleinberg, J.: Small-world phenomena and the dynamics of information. In: NIPS 2001. Advances in Neural Information Processing Systems (2001)
9. Kleinberg, J.: Complex networks and decentralized search algorithms. In: ICM 2006. Proc. the International Congress of Mathematicians (2006)
10. Leskovec, J., Chakrabarti, D., Kleinberg, J., Faloutsos, C.: Realistic, mathematically tractable graph generation and evolution, using kronecker multiplication. In: Jorge, A.M., Torgo, L., Brazdil, P.B., Camacho, R., Gama, J. (eds.) PKDD 2005. LNCS (LNAI), vol. 3721, Springer, Heidelberg (2005)
11. Leskovec, J., Faloutsos, C.: Scalable modeling of real graphs using kronecker multiplication. In: ICML 2007. International Conference on Machine Learning (2007)
12. Leskovec, J., Kleinberg, J., Faloutsos, C.: Graph evolution: Densification and shrinking diameters. ACM Transactions on Knowledge Discovery from Data 1(1) (2007)
13. Mahdian, M., Xu, Y.: Stochastic kronecker graphs. Technical Report (2007)

Deterministic Decentralized Search
in Random Graphs

Esteban Arcaute[1,*], Ning Chen[2,*], Ravi Kumar[3], David Liben-Nowell[4,*],
Mohammad Mahdian[3], Hamid Nazerzadeh[1,*], and Ying Xu[1,*]

[1] Stanford University
{arcaute,hamidnz,xuying}@stanford.edu
[2] University of Washington
ning@cs.washington.edu
[3] Yahoo! Research
{ravikuma,mahdian}@yahoo-inc.com
[4] Carleton College
dlibenno@carleton.edu

Abstract. We study a general framework for decentralized search in random graphs. Our main focus is on deterministic memoryless search algorithms that use only local information to reach their destination in a bounded number of steps in expectation. This class includes (with small modifications) the search algorithms used in Kleinberg's pioneering work on long-range percolation graphs and hierarchical network models. We give a characterization of searchable graphs in this model, and use this characterization to prove a monotonicity property for searchability.

1   Introduction

Since Milgram's famous "small world" experiment [14], it has generally been understood that social networks have the property that a typical node can reach any other node through a short path (the so-called "six degrees of separation"). An implication of this fact is that social networks have small diameter. Many random graph models have been proposed to explain this phenomenon, often by showing that adding a small number of random edges causes a highly structured graph to have a small diameter (e.g., [16,3]). A stronger implication of Milgram's experiment, as Kleinberg observed [8], is that for most social networks there are decentralized search algorithms that can *find* a short path from a source to a destination without a global knowledge of the graph. As Kleinberg proved, even many of the random graph models with small diameter do not have this property (i.e., any decentralized search algorithm in such graphs can take many steps to reach the destination), while in certain graph models with a delicate balance of parameters, decentralized search is possible. Since Kleinberg's work, there have been many other models that provably exhibit the searchability property [10,7,15,12,9,5]; however, we still lack a good understanding of what contributes to this property in graphs.

---

* Work performed in part while visiting Yahoo! Research.

A. Bonato and F.R.K. Chung (Eds.): WAW 2007, LNCS 4863, pp. 187–194, 2007.

In this paper, we look at a general framework for searchability in random graphs. We consider a general random graph model in which the set of edges leaving a node u is independent of that of any other node $v \neq u$. This framework includes models such as the directed variant of the classical Erdős–Rényi graphs [6], random graphs with a given expected degree sequence (e.g., [4]), long-range percolation graphs [8], hierarchical network models [9], and graphs based on Kronecker products [11,13], but not models such as preferential attachment [2] in which the distribution of edges leaving a node is dependent on the other edges of the graph. It is worth noting that, in a random graph model where edges can have arbitrary dependencies, the search problem includes arbitrarily difficult learning problems as special cases, and therefore one cannot expect to have a complete characterization of searchable graphs in such a model.

Throughout most of this paper, we restrict the class of decentralized search algorithms that we consider to deterministic memoryless algorithms that succeed in finding a path to the destination with probability 1. This is an important class of search algorithms, and includes the decentralized search algorithms used in Kleinberg's work on long-range percolation graphs and hierarchical network models. For this class, we give a simple characterization of graphs that are searchable in terms of a node ordering property. We will use this characterization to show a monotonicity property for searchability: if a graph is searchable in our model, it stays searchable if the probabilities of edges are increased.

The rest of this paper is organized as follows: Section 2 contains the description of the model. Section 3 presents a characterization of searchable random graphs. The monotonicity theorem is presented in Section 4.

2   The Model

Given a positive integer n and an $n \times n$ matrix $\mathbf{P}$ with entries $p_{i,j} \in [0,1]$, we define a directed random graph $G(n, \mathbf{P})$ with the node set $V = \{1, \ldots, n\}$ and with a directed edge connecting node i to node j with probability p_{ij}, independently of all other edges. As we will see later (Remark 1), our results hold for a more general random graph model where the edges originating from a node i can be dependent on each other but are independent of the edges leaving other nodes. However, for the sake of simplicity, we state and prove our results in the $G(n, \mathbf{P})$ model.

We fix two nodes $s, t \in V$ of $G(n, \mathbf{P})$ as the *source* and the *destination*. For $v \in V$, let $\Gamma(v)$ denote the set of out-neighbors of u in G. We investigate the existence of a decentralized search algorithm that finds a path from s to t of at most a given length d in expectation.[1] We restrict our attention to *deterministic memoryless algorithms*. A deterministic memoryless algorithm can be defined

---

[1] Alternatively, we could ask for which graphs a decentralized search algorithm can find a path between *every* pair of nodes s and t, or between a *random* pair of nodes s and t. Our techniques apply to these alternative formulations of the problem as well. The only point that requires some care is that the orderings in the characterization theorem can depend on s and t.

as a partial function $A : V \times 2^V \to V$. Such an algorithm A defines a path $v_0, v_1, v_2, \ldots$ on a given graph G as follows: $v_0 = s$, and for every $i \geq 0$, $v_{i+1} = A(v_i, \Gamma(v_i))$. The length of this path is defined as the smallest integer i such that $v_i = t$. If no such i exists, we define the length of the path as infinity.

We are now ready to define the notion of searchability. For a given matrix $\mathbf{P}$, source and destination nodes s and t, and a number d, we say that $G(n, \mathbf{P})$ is d-searchable using a deterministic memoryless algorithm A if the expected length of the path defined by A on $G(n, \mathbf{P})$ is at most d. Note that this definition requires that the algorithm find a path from s to t with probability 1.

3 A Characterization of Searchable Random Graphs

In this section, we provide a complete characterization of searchable random graphs. We begin by defining a class of deterministic memoryless search algorithms parameterized by two orderings of V, and then prove that if a graph is d-searchable, it is also d-searchable using an algorithm from this narrow class.

Definition 1. *Let σ, π be two orderings (i.e., permutations) of the node set V. We define a deterministic memoryless algorithm $A_{\sigma,\pi}$ corresponding to these orderings as follows: for every $u \in V$, $A_{\sigma,\pi}(u, \Gamma(u))$ is defined as the maximum element according to π of the set $\{v \in \Gamma(u) : \sigma(v) > \sigma(u)\} \cup \{u\}$.*

In other words, algorithm $A_{\sigma,\pi}$ never goes backwards according to the ordering σ, and, subject to this restriction, makes the maximum possible progress according to π.

We are now ready to state our main theorem.

Theorem 1. *For a given probability matrix $\mathbf{P}$, source and destination nodes s and t, and number d, if $G(n, \mathbf{P})$ is d-searchable using a deterministic memoryless algorithm A, then there exist two orderings σ and π of V such that $G(n, \mathbf{P})$ is d-searchable by using $A_{\sigma,\pi}$.*

To prove this theorem, we will first construct the ordering σ using the structure of the search algorithm A. Next, we define an ordering π using σ. Finally, we use induction with respect to the ordering σ to show that the expected length of the path defined by $A_{\sigma,\pi}$ on $G(n, \mathbf{P})$ is not more than the one defined by A.

We assume, without loss of generality, that for every set $S \subseteq V$, $A(t, S) = t$. In other words, we assume that A never leaves t once it reaches this node.

Define a graph H with the node set V as follows: for every pair $u, v \in V$, the edge (u, v) is in H if and only if this edge is on the path from s to t defined by A on some realization of $G(n, \mathbf{P})$ (i.e., on some graph that has a non-zero probability in the distribution $G(n, \mathbf{P})$). We have the following important lemma.

Lemma 1. *The graph H is acyclic.*

Proof. Assume, for contradiction, that H contains a simple cycle C. Note that by the definition of H, if an edge (u, v) is in H, then u must be reachable from s

in H. Therefore, every node of C must be reachable from s in H. Let v^* be a node in C that has the shortest distance from s in H, and $s = v_0, v_1, \ldots, v_\ell = v^*$ be a shortest path from s to v^* in H. Also, let $v^* = v_\ell, v_{\ell+1}, \ldots, v_k, v_{k+1} = v^*$ denote the cycle C. Therefore, $v_0, v_1, \ldots, v_k$ are all distinct nodes, and for every $i \in \{0, \ldots, k\}$, there is an edge from v_i to v_{i+1} in H.

By the definition of H, for every $i \in \{0, \ldots, k\}$, there is a realization of $G(n, \mathbf{P})$ in which A traverses the edge (v_i, v_{i+1}). This means that there is a realization of $G(n, \mathbf{P})$ in which the set $\Gamma(v_i)$ of out-neighbors of v_i is S_i^*, for some set S_i^* such that $A(v_i, S_i^*) = v_{i+1}$. Recall that in $G(n, \mathbf{P})$, all edges are present independently at random, and thus the random variables $\Gamma(u)$ are independent. Hence, since v_i's are all distinct and for each i, there is a realization satisfying $\Gamma(v_i) = S_i^*$, there must be a realization in which $\Gamma(v_i) = S_i^*$ for *all* i. In this realization, the algorithm A falls in the cycle C, and therefore will never reach t. Thus the path found by A in this realization is infinitely long, and therefore the expected length of the path found by A is infinite. This is a contradiction. □

By Lemma 1, we can find a topological ordering of the graph H. Furthermore, since by assumption t has no outgoing edge in H, we can find a topological ordering that places t last. Let σ be such an ordering, i.e., σ is an ordering of V such that (i) t is the maximum element of V under σ; (ii) for every edge (u, v) in H, we have $\sigma(v) > \sigma(u)$; and (iii) all nodes not in H precede s and are ordered arbitrarily, i.e., $\sigma(s) > \sigma(v)$ for any such node v. By the definition of H, these conditions mean that the algorithm A (starting from the node s) never traverses an edge (u, v) with $\sigma(u) > \sigma(v)$.

Given the ordering σ, we define numbers r_u for every $u \in V$ recursively as follows: $r_t = 0$, and for every $u \neq t$,

$$
r_u = \begin{cases} 1 + \displaystyle\sum_{S \subseteq T_u, S \neq \emptyset} q_{u,S} \cdot \min_{v \in S}\{r_v\} & \text{if } q_{u,\emptyset} = 0 \\ \infty & \text{if } q_{u,\emptyset} > 0, \end{cases} \tag{1}
$$

where $T_u := \{v : \sigma(v) > \sigma(u)\}$ and, for a set $S \subseteq T_u$, we write

$$
q_{u,S} := \left(\prod_{v \in S} p_{uv} \right) \left(\prod_{v \in T_u \setminus S} (1 - p_{uv}) \right)
$$

to denote the probability that the subset of nodes of T_u that are out-neighbors of u is precisely S. Note that the above formula defines r_u in terms of r_v for $\sigma(v) > \sigma(u)$, and therefore the definition is well founded.

We can now define the ordering π as follows: let $\pi(u) > \pi(v)$ if $r_u < r_v$. Pairs u, v with $r_u = r_v$ are ordered arbitrarily by π.

The final step of the proof is the following lemma, which we will prove by induction using the ordering σ. To state the lemma, we need a few pieces of notation. For a search algorithm B, let $d(B, u)$ denote the expected length of the path that the algorithm B, started at node u, finds to t. Also, let V_0 denote the set of non-isolated nodes of H—i.e., V_0 is the set of nodes that the algorithm A (started from s) has a non-zero chance of reaching.

Lemma 2. *Let σ and π be the orderings defined as above. Then for every node $u \in V_0$, we have that $d(A, u) \geq d(A_{\sigma,\pi}, u) = r_u$.*

Proof (sketch). We prove this statement by induction on u, according to the ordering σ. The statement is trivial for $u = t$. We now show that for $u \in V_0 \setminus \{t\}$ if the statement holds for every node $v \in V_0$ with $\sigma(v) > \sigma(u)$, then it also holds for u. Observe that for any deterministic memoryless algorithm B,

$$d(B, u) = 1 + \sum_{S \subset V, S \neq \emptyset} q'_{u,S} \cdot d(B, B(u, S)), \tag{2}$$

where $q'_{u,S} := (\prod_{v \in S} p_{uv})(\prod_{v \in V \setminus S}(1 - p_{uv}))$ is the probability that the set of out-neighbors of u in $G(n, \mathbf{P})$ is precisely S. This statement follows from the fact that the algorithm B is memoryless, and the fact that $q'_{u,\emptyset} = 0$ since $u \in V_0$. Applying Equation (2) to $A_{\sigma,\pi}$ and using the fact that, by definition, $A_{\sigma,\pi}(u, S)$ only depends on u and $S \cap T_u$, we obtain

$$d(A_{\sigma,\pi}, u) = 1 + \sum_{S \subseteq T_u, S \neq \emptyset} q_{u,S} \cdot d(A_{\sigma,\pi}, A_{\sigma,\pi}(u, S)). \tag{3}$$

We have that $d(A_{\sigma,\pi}, A_{\sigma,\pi}(u, S)) = r_{A_{\sigma,\pi}(u,S)}$ by the induction hypothesis. Also, by the definition of $A_{\sigma,\pi}$ and π, we have that $r_{A_{\sigma,\pi}(u,S)} = \min_{v \in S}\{r_v\}$. Combined with Equation (3) and the definition of r_u, this shows $d(A_{\sigma,\pi}, u) = r_u$, as desired.

To prove $d(A, u) \geq r_u$, note that since $A(u, S) \in S \cap T_u \cap V_0$, we have

$$d(A, A(u, S)) \geq \min_{v \in S \cap T_u \cap V_0}\{d(A, v)\}.$$

By the induction hypothesis, we have that $d(A, v) \geq r_v$ for every $v \in T_v \cap V_0$. Therefore, we have that $d(A, A(u, S)) \geq \min_{v \in S \cap T_u \cap V_0}\{r_v\}$. Substituting this in Equation (2), we obtain

$$d(A, u) \geq 1 + \sum_{S \subset V, S \neq \emptyset} q'_{u,S} \cdot \min_{v \in S \cap T_u \cap V_0}\{r_v\}$$

$$= 1 + \sum_{S \subseteq T_u, S \neq \emptyset} q_{u,S} \cdot \min_{v \in S \cap T_u \cap V_0}\{r_v\}$$

$$\geq r_u.$$

This completes the proof of the induction step. □

Proof (of Theorem 1). Define the graph H, the ordering σ, the values r_u, and the ordering π as above. By Lemma 2, we have that $d(A_{\sigma,\pi}, s) \leq d(A, s)$. Since $G(n, \mathbf{P})$ is d-searchable using A by assumption, we have that $d(A, s) \leq d$. Hence we have $d(A_{\sigma,\pi}, s) \leq d$, as desired. □

Remark 1. It is not hard to see that the only property of $G(n, \mathbf{P})$ that was used in the above proof was the fact that the random variables $\Gamma(u)$ (the set of out-neighbors of u) are independent. Therefore, the above proof (with minor modifications in the definitions of $q_{u,S}$ and $q'_{u,S}$) also works for a more general model of random graphs. This includes the directed ACL graphs [1] and the long-range percolation graphs.

Note that in the above proof, the second ordering π was defined in terms of the first ordering σ and $\mathbf{P}$. Therefore, the condition for the searchability of $G(n, \mathbf{P})$ can be stated in terms of only one ordering σ as follows:

Corollary 2. *$G(n, \mathbf{P})$ is d-searchable if and only if there is an ordering σ on the nodes for which $r_s \leq d$, where r is defined as in (1).*

It is not hard to see that even though the expression on the right-hand side of (1) has exponentially many terms, given σ, the value of r_u can be computed in polynomial time for every u. Therefore, the above corollary reduces the problem of d-searchability of $G(n, \mathbf{P})$ to a node-ordering property with a tractable objective function.

4   The Monotonicity Property

Armed with the characterization theorem of the previous section, we can now prove the following natural monotonicity property for searchability.

Theorem 3. *Let $\mathbf{P}$, $\mathbf{P}'$ be two $n \times n$ probability matrices such that for every i and j, we have $p_{ij} \leq p'_{ij}$. Fix the source and destination nodes s and t. Then, if $G(n, \mathbf{P})$ is d-searchable for some d, so is $G(n, \mathbf{P}')$.*

Proof (sketch). By Corollary 2, since $G(n, \mathbf{P})$ is d-searchable, there is an ordering σ such that the value r_s defined using Equation (1) is at most d. To show d-searchability of $G(n, \mathbf{P}')$, we apply the same ordering σ. Let $\{r'_u\}$ denote the values computed using Equation (1), but with $\mathbf{P}$ replaced by $\mathbf{P}'$. All we need to do is to show that $r'_s \leq d$ and then use Corollary 2. To do this, we prove by induction that for every $u \in V_0$, we have $r'_u \leq r_u$. This statement is trivial for $u = t$. We assume it is proved for every $v \in V_0$ with $\sigma(v) > \sigma(u)$, and prove it for u.

We have

$$r'_u = 1 + \sum_{S \subseteq T_u, S \neq \emptyset} \prod_{v \in S} p'_{uv} \prod_{v \in T_u \setminus S} (1 - p'_{uv}) \cdot \min_{v \in S} \{r'_v\}$$

$$\leq 1 + \sum_{S \subseteq T_u, S \neq \emptyset} \prod_{v \in S} p'_{uv} \prod_{v \in T_u \setminus S} (1 - p'_{uv}) \cdot \min_{v \in S} \{r_v\}$$

Let $1, 2, \ldots, k$ be the nodes of T_u, ordered in such a way that $r_1 \geq r_2 \geq \cdots \geq r_k$. It is not hard to see that the above expression can be written as follows.

$$r'_u \leq 1 + r_1 - \sum_{i=1}^{k-1} \mathbf{Pr}_{G(n,\mathbf{P}')}[\Gamma(u) \cap \{i+1, \ldots, k\} \neq \emptyset] \cdot (r_i - r_{i+1})$$

The coefficient of $(r_i - r_{i+1})$ in the above expression is the probability of the event that the set of nodes that have an edge from u in $G(n, \mathbf{P}')$ contains at least one of the nodes $i+1, \ldots, k$. This event is monotone; therefore the probability of this event under $G(n, \mathbf{P})$ is less than or equal to the probability under $G(n, \mathbf{P}')$. Therefore,

$$r'_u \leq 1 + r_1 - \sum_{i=1}^{k-1} \mathbf{Pr}_{G(n,\mathbf{P})}[\Gamma(u) \cap \{i+1, \ldots, k\} \neq \emptyset] \cdot (r_i - r_{i+1}).$$

This completes the proof of the induction step, since the right-hand side of the above inequality is precisely r_u. $\qquad\square$

Note that, simple as the statement of Theorem 3 sounds, we do not know whether a similar statement holds for randomized memoryless algorithms. On the other hand, we proved the monotonicity property for randomized algorithms with memory; the proof can be found in [13].

Acknowledgments

We thank Amin Saberi for many useful discussions.

References

1. Aiello, W., Chung, F., Lu, L.: A random graph model for power law graphs. Experimental Mathematics 10(1), 53–66 (2001)
2. Barabási, A.-L., Albert, R.: Emergence of scaling in random networks. Science 286(15), 509–512 (1999)
3. Bollobás, B., Chung, F.R.K.: The diameter of a cycle plus a random matching. SIAM Journal on Discrete Mathematics 1(3), 328–333 (1988)
4. Chung, F., Lu, L.: The average distance in a random graph with given expected degrees. Internet Mathematics 1(1), 91–114 (2003)
5. Duchon, P., Hanusse, N., Lebhar, E., Schabanel, N.: Could any graph be turned into a small world? Theoretical Computer Science 355(1), 96–103 (2006)
6. Erdős, P., Rényi, A.: On random graphs I. Publications Mathematics Debrecen 6, 290–297 (1959)
7. Fraigniaud, P.: Greedy routing in tree-decomposed graphs. In: Proceedings of the 13th Annual European Symposium on Algorithms, pp. 791–802 (2005)
8. Kleinberg, J.: The small-world phenomenon: An algorithmic perspective. In: Proceedings of the 32nd ACM Symposium on Theory of Computing, pp. 163–170 (2000)
9. Kleinberg, J.: Small-world phenomena and the dynamics of information. Advances in Neural Information Processing Systems 14, 431–438 (2001)
10. Kumar, R., Liben-Nowell, D., Tomkins, A.: Navigating low-dimensional and hierarchical population networks. In: Proceedings of the 14th Annual European Symposium on Algorithms, pp. 480–491 (2006)

11. Leskovec, J., Chakrabarti, D., Kleinberg, J., Faloutsos, C.: Realistic, mathematically tractable graph generation and evolution, using Kronecker multiplication. In: Proceedings of the 10th European Conference on Principles and Practice of Knowledge Discovery in Databases, pp. 133–145 (2005)
12. Liben-Nowell, D., Novak, J., Kumar, R., Raghavan, P., Tomkins, A.: Geographic routing in social networks. Proceedings of the National Academy of Sciences 102(33), 11623–11628 (2005)
13. Mahdian, M., Xu, Y.: Stochastic Kronecker graphs. In: Proceedings of the 5th Workshop on Algorithms and Models for the Web-Graph (2007)
14. Milgram, S.: The small world problem. Psychology Today 1, 61–67 (1967)
15. Slivkins, A.: Distance estimation and object location via rings of neighbors. In: Proceedings of the 16th ACM Symposium on Principles of Distributed Computing, pp. 41–50 (2005)
16. Watts, D.J., Strogatz, S.H.: Collective dynamics of small-world networks. Nature 393, 440–442 (1998)

Using Bloom Filters to Speed Up HITS-Like Ranking Algorithms

Sreenivas Gollapudi, Marc Najork, and Rina Panigrahy

Microsoft Research, Mountain View CA 94043, USA

Abstract. This paper describes a technique for reducing the query-time cost of HITS-like ranking algorithm. The basic idea is to compute for each node in the web graph a summary of its immediate neighborhood (which is a query-independent operation and thus can be done off-line), and to approximate the neighborhood graph of a result set at query-time by combining the summaries of the result set nodes. This approximation of the query-specific neighborhood graph can then be used to perform query-dependent link-based ranking algorithms such as HITS and SALSA. We have evaluated our technique on a large web graph and a substantial set of queries with partially judged results, and found that its effectiveness (retrieval performance) is comparable to the original SALSA algorithm, while its efficiency (query-time speed) is substantially higher.

1 Introduction

One of the fundamental problems in Information Retrieval is the *ranking problem*: how to arrange the documents that satisfy a query into an order such that the documents most relevant to the query rank first. Traditional ranking algorithms proposed by the pre-web IR community were mostly based on similarity measures between the terms (words) in the query and the documents satisfying the query.

In addition to structured text, web pages also contain hyperlinks between web pages, which can be thought of as peer endorsements between content providers. Marchiori suggested early on to leverage incoming hyperlinks as another feature in ranking algorithms [9], and the simplistic idea of merely counting in-links quickly evolved into more sophisticated link-based ranking algorithms that take the quality of an endorsing web page into account.

Link-based ranking algorithms can be grouped into two classes: query-independent ones such as in-link count or Google's famous PageRank [12], and query-dependent ones such as Kleinberg's HITS [4,5] and Lempel & Moran's SALSA [6,7]. The aforementioned algorithms were described in seminal papers that inspired a great deal of subsequent work; however, there has been little published work on the effectiveness (that is, the accuracy of the ranking) of these algorithms.

A recent study using a 17-billion edge web graph and a set of 28,043 queries with partially judged results concluded that SALSA, a query-dependent link-based ranking algorithm, is substantially more effective than HITS, PageRank

A. Bonato and F.R.K. Chung (Eds.): WAW 2007, LNCS 4863, pp. 195–201, 2007.

and in-degree, although it is not as effective as the state-of-the-art textual ranking algorithm.

Unfortunately, SALSA in particular and HITS-like algorithms in general require a substantial amount of expense at query-time. The vast fraction of this expense is devoted to determining the *neighborhood graph* of the results to a query; the subsequent computation of scores for the nodes in the neighborhood graph is cheap in comparison, despite the fact that most HITS-like algorithms use power iteration to compute the fixed-points of the score vectors. The fact that HITS-like algorithms incur substantial computational cost at query-time puts them at a crippling disadvantage to query-independent algorithms such as PageRank: according to Marissa Mayer, Google's VP of Search Products & User Experience, delaying Google's response time by half a second led to a 20% drop in query traffic (and revenue) from the user population subjected to the delay [8].

This paper describes a technique that dramatically lowers the query-time cost of HITS-like ranking algorithms, *i.e.* algorithms that perform computations on the distance-one neighborhood graph of the results to a query. The basic idea is to compute a *summary* of the neighborhood of each page on the web (an operation that is query-independent and thus can be done off-line, say at index construction time), and to use these summaries at query time to *approximate* the neighborhood graph of the result set and to compute scores using the approximate graph. We have evaluated this approach using the same methodology and the same data sets that were used in the earlier comparisons of in-degree, PageRank, HITS, and SALSA, and found that applying our technique to SALSA does not impair its effectiveness and at the same substantially improves its efficiency.

The remainder of this paper is structured as follows: section 2 reviews the HITS and SALSA algorithms; section 3 explains our technique for summarizing the neighborhood of each web page; section 4 describes the experimental validation of our technique; and section 5 offers concluding remarks.

2   HITS and SALSA

Both HITS and SALSA are query-dependent link-based ranking algorithms: given a web graph (V, E) with vertex set V and edge set $E \subseteq V \times V$ (where edges/links between vertices/pages on the same web server are typically omitted), and the set of result URLs to a query (called the *root set $R \subseteq V$*) as input, both compute a *base set $B \subseteq V$*, defined to be:

$$B = R \cup \bigcup_{u \in R} \{v \in V : (u, v) \in E\} \cup \bigcup_{v \in R} \mathcal{S}_n[\{u \in V : (u, v) \in E\}]$$

where $\mathcal{S}_n[X]$ denotes a uniform random sample of n elements from set X ($\mathcal{S}_n[X] = X$ if $|X| < n$). The sampling parameter n will typically be below 100, and its choice has a significant impact on the effectiveness of SALSA in particular [11]. The neighborhood graph (B, N) consists of base set B and edge set $N = \{(u, v) \in E : u \in B \land v \in B\}$.

Both HITS and SALSA compute two scores for each node $u \in B$: an authority score $A(u)$, estimating how authoritative u is on the topic induced by the query, and a hub score $H(u)$, indicating whether u is a good reference to many authoritative pages. In the case of HITS, hub scores and authority scores are computed in a mutually recursive fashion:

1. For all $u \in B$ do $H(u) := \sqrt{\frac{1}{|B|}}, A(u) := \sqrt{\frac{1}{|B|}}$.
2. Repeat until H and A converge:
   (a) For all $v \in B$ do $A'(v) := \sum_{(u,v) \in N} H(u)$
   (b) For all $u \in B$ do $H'(u) := \sum_{(u,v) \in N} A(v)$
   (c) For all $u \in B$ do $H(u) := \frac{1}{\|H'\|_2} H'(u), A(u) := \frac{1}{\|A'\|_2} A'(u)$

where $\|X\|_2$ is the euclidean norm of vector X.

By contrast, SALSA authority scores can be computed independently of hub scores, which is interesting insofar as that SALSA (and HITS) hub scores are poor ranking features. The algorithm for computing SALSA authority scores is:

1. Let B^A be $\{u \in B : in(u) > 0\}$
2. For all $u \in B$ do $A(u) := \begin{cases} \frac{1}{|B^A|} & \text{if } u \in B^A \\ 0 & \text{otherwise} \end{cases}$
3. Repeat until A converges:
   (a) For all $u \in B^A$ do $A'(u) := \sum_{(v,u) \in N} \sum_{(v,w) \in N} \frac{A(w)}{out(v)in(w)}$
   (b) For all $u \in B^A$ do $A(u) := A'(u)$

For reasons of space, we omit the algorithm for computing SALSA hub scores; the interested reader is referred to [6] or [11]. The latter paper compares the effectiveness of HITS and SALSA, and finds that SALSA authority scores are a substantially better ranking feature than HITS authority scores. Our experimental validation in section 4 employs the same methodology and data sets.

When performed on a web-scale corpus, both HITS and SALSA require a substantial amount of query time processing. Much of this processing is attributable to the computation of the neighborhood graph. The reason for this is that the entire web graph is enormous. A document collection of five billion web pages induces a set of about a quarter of a trillion hyperlinks. Storing this web graph on disk would make lookup operations unacceptably slow due to the inherent seek time limitations of hard drives. On the other hand, the graph is too big to be stored in the main memory of any single machine; therefore, it has to be partitioned across many machines. In such a setup, the cost of a link lookup is governed by the cost of a remote procedure call (RPC). A sophisticated implementation of the SALSA algorithm against a distributed link database will batch many lookup operations into a single RPC request to reduce latency and will query all link servers in parallel, but even so it will require four rounds of concurrent requests: one round to map the root set URLs to short representations; the second round to determine the pages linking to the root set; the third round to determine the pages pointed to by the root set; and the fourth round to determine the edges induced by the base set. The SALSA implementation used to perform the experiments described in [11] required 235 milliseconds per query for the most effective parametrization of SALSA, and 2.15 seconds per query for the most expensive parametrization. Over 90% of the time spent was spent on performing the RPC calls to the link servers in order to assemble the neighborhood graph, as opposed to computing scores on that graph.

3   Summarizing Neighborhood Graphs

In this paper, we present a technique to substantially lower the query-time cost of HITS and SALSA. We do so by moving the most expensive part of the computation off-line. At index-construction time, we build a database mapping web page URLs to summaries of their neighborhoods. At query time, we rank the results satisfying a query by looking up each result in the summary database (an operation that requires only one round of RPCs, as opposed to four), approximating the neighborhood graph of the result set based on the neighborhood summaries of the constituent results, and computing SALSA (or HITS) scores using that approximation of the neighborhood graph. In the experimental evaluation below, we will show that this approximation has no detrimental effect on the quality of the ranking.

As outlined above, our summary of the neighborhood graph of a web page u consists of a summary of the *ancestors* (the pages that link to u) and a summary of the *descendants* (the pages that u links to), each consisting of a Bloom filter containing a limited-size subset of ancestors or descendants plus a much smaller subset containing explicit web page identifiers (64-bit integers). A Bloom filter is a space-efficient probabilistic data structure that can be used to test the membership of an element in a given set; the test may yield a false positive but never a false negative. A Bloom filter represents a set using an array A of m bits (where $A[i]$ denotes the ith bit), and uses k hash functions h_1 to h_k to manipulate the array, each h_i mapping some element of the set to a value in $[1,m]$. To add an element e to the set, we set $A[h_i(e)]$ to 1 for each $1 \leq i \leq k$; to test whether e is in the set, we verify that $A[h_i(e)]$ is 1 for all $1 \leq i \leq k$. Given a Bloom filter size m and a set size n, the optimal (false-positive minimizing) number of hash functions k is $\frac{m}{n} \ln 2$; in this case, the probability of false positives is $(\frac{1}{2})^k$. For an in-depth description of Bloom filters, the reader is referred to [1,3]. In the following, we will write $BF[X]$ to denote the Bloom filter representing the set X.

While the original SALSA algorithm samples the neighborhood (and in particular the ancestors) uniformly *at random*, we use consistent sampling [2]. Let $\mathcal{C}_n[X]$ denote a consistent unbiased sample of n elements from set X, where $\mathcal{C}_n[X] = X$ if $|X| < n$. Consistent sampling is *deterministic*; that is, when sampling n elements from a set X, we will always draw the same n elements. Moreover, any element x that is sampled from set A is also sampled from subset $B \subset A$ if $x \in B$. We define the set $I_n(u)$ to be a consistent sample $\mathcal{C}_n[\{v \in V : (v,u) \in E\}]$ of (at most) n of the ancestors of u; and the set $O_n(u)$ to be a consistent sample $\mathcal{C}_n[\{v \in V : (u,v) \in E\}]$ of n of the descendants of u. For each page u in the web graph, we compute a summary $(BF[I_a(u)], I_b(u), BF[O_c(u)], O_d(u))$.

At query time, given a result set R, we first look up the summaries for all the results in R. Having done so, we compute a cover set

$$C = R \cup \bigcup_{u \in R} I_b(u) \cup \bigcup_{u \in R} O_d(u)$$

Next, we construct a graph consisting of the vertices in C. We fill in the edges as follows: For each vertex $u \in R$ and each vertex $v \in C$, we perform two tests: If $BF[I_a(u)]$ contains v, we add an edge (v, u) to the graph; if $BF[O_c(u)]$ contains v, we add an edge (u, v) to the graph. This graph serves as our approximation of the neighborhood graph of R; we use it to compute SALSA (or HITS) scores using the same algorithm as described above in section 2.

Observe that the approximate neighborhood graph differs from the exact neighborhood graph in three ways:

- In the exact graph, we do not sample the vertices directly reachable from the root set, but rather include them all.
- The approximate graph only contains edges from $C \cap I_a(u)$ to $u \in R$ and from $u \in R$ to $C \cap O_c(u)$. In other words, it excludes edges between nodes in C that are not part of the root set.
- We do not use exact set representations for $I_a(u)$ and $O_c(u)$, but approximate them by using Bloom filters. This introduces additional edges whose number depends on the false positive probability of the Bloom filter. Using k hash functions, we add about $2^{-k+1}|C||R|$ spurious edges in the graph.

At first glance, it is non-obvious why this approximation of the neighborhood graph should preserve any of the properties relevant to ranking algorithms. After all, the approximation may exclude actual edges due to the sampling process, and add phantom edges due to the potential for false positives inherent to Bloom filters. However, it is worth noting that consistent sampling preserves co-citation relationships between pages in the result set. The experimental validation described in the following section confirms that the summarization algorithm indeed preserves properties relevant to link-based ranking.

4   Experimental Validation

Our experimental validation is based on the two data sets used in [10,11]: a large web graph with 2.9 billion nodes and 17.7 billion edges, and a set of 28,043 queries, each with 2,838 results on average, 17.3 of which were rated by human judges on a six-point scale. For more details on how the graph was collected and how results were judged, the reader is referred to the earlier papers. We use three popular measures of effectiveness (or *retrieval performance*): the *normalized discounted cumulative gain* (NDCG), the *mean average precision* (MAP), and the *mean reciprocal rank* (MRR) measures. All three measures are normalized

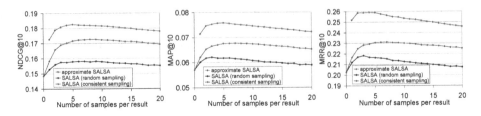

Fig. 1. Effectiveness of authority scores computed using different parameterizations of original and approximate SALSA; measured in terms of NDCG, MAP and MRR

to range between 0 and 1; higher values indicate better performance. Again, the reader is referred to [10,11] for the full definitions of these measures.

In our experiments, we used $k = 10$ hash functions, and we fixed the parameters a and c at 1000; that is, we included a sample of (up to) 1000 ancestors or descendants into each Bloom filter. The Bloom filters averaged 227 bytes for ancestor sets and 72 bytes for descendant sets. We measured the effectiveness of SALSA using our summarization technique for b and d ranging from 2 to 20, and compared it to original SALSA with the same sampling values. Figure 1 depicts the results. The figure contains three graphs, one for each performance measure (NDCG, MAP, and MRR). The horizonal axis shows the number of samples (the b and d parameters of our summarization-based SALSA, and the n parameter of the original SALSA); the vertical axis shows the retrieval performance. Each graph contains three curves; the blue (dark) curve showing the performance of the original SALSA; the green (medium) curve shows the performance of the original SALSA with consistent instead of random sampling; and the red (light) curve showing that of our summarization-based version. Using consistent instead of random sampling substantially improves the performance under all measures. However, our new approximate version of SALSA outperforms the original SALSA algorithm under all measures. Performance is maximal for b and d between 4 and 5, depending on the measure. For $b = d = 5$, each summary is 379 bytes in size (227 bytes for $BF[I_{1000}]$, 40 bytes for I_5, 72 bytes for $BF[O_{1000}]$, and 40 bytes for O_5).

Our current implementation of summarization-based SALSA does not yet employ a distributed summary server; we use our distributed link server to compute summaries. However, since a summary server is similar to, and indeed simpler than, a link server, we can draw on our experiences with the latter to estimate what the query-time performance of a production system would be. We measured the performance of our current implementation, subtracted the time required to compute the summaries, and added the time we have observed for retrieving a vector of links of the size of an average summary from our link server. These measurements suggest that it would take us an average of 171 milliseconds per query to compute approximate SALSA scores. This compares favorably to the 235 milliseconds per query of our implementation of the original SALSA algorithm. Moreover, we have not spent any time on optimizing the code for constructing approximate neighborhood graphs, and believe that further speedups are possible.

5 Conclusion

This paper describes a technique for reducing the query-time cost of query-dependent link-based ranking algorithms that operate on the distance-one neighborhood graph of the result set, such as HITS and SALSA. Our technique computes a summary of each page on the web ahead of query-time, and combines these summaries at query-time into approximations of the neighborhood graph of the result set. We tested our technique by applying it to a large real web graph and a sizable collection of real queries with partially assessed results, and were able to demonstrate that the approximate nature of our technique does not have any negative impact on the effectiveness (retrieval performance). Future work includes implementing a distributed summary server and verifying that our technique is indeed faster than the original (all-online) implementation.

References

1. Bloom, B.: Space/time trade-offs in hash coding with allowable errors. Communications of the ACM 13(7), 422–426 (1970)
2. Broder, A., Charikar, M., Frieze, A., Mitzenmacher, M.: Min-Wise Independent Permutations. Journal of Computer and System Sciences 60(3), 630–659 (2000)
3. Broder, A., Mitzenmacher, M.: Network Applications of Bloom Filters: A Survey. Internet Mathematics 1(4), 485–509
4. Kleinberg, J.M.: Authoritative sources in a hyperlinked environment. In: Proc. of the 9th Annual ACM-SIAM Symposium on Discrete Algorithms, pp. 668–677 (1998)
5. Kleinberg, J.M.: Authoritative sources in a hyperlinked environment. Journal of the ACM 46(5), 604–632 (1999)
6. Lempel, R., Moran, S.: The stochastic approach for link-structure analysis (SALSA) and the TKC effect. Computer Networks and ISDN Systems 33(1-6), 387–401 (2000)
7. Lempel, R., Moran, S.: SALSA: The stochastic approach for link-structure analysis. ACM Transactions on Information Systems 19(2), 131–160 (2001)
8. Linden, G.: Marissa Mayer at Web 2.0, Online at
http://glinden.blogspot.com/2006/11/marissa-mayer-at-web-20.html
9. Marchiori, M.: The quest for correct information on the Web: Hyper search engines. Computer Networks and ISDN Systems 29(8-13), 1225–1236 (1997)
10. Najork, M., Zaragoza, H., Taylor, M.: HITS on the Web: How does it Compare? In: 30th Annual International ACM SIGIR Conference on Research and Development in Information Retrieval, pp. 471–478 (2007)
11. Najork, M.: Comparing the Effectiveness of HITS and SALSA. In: 16th ACM Conference on Information and Knowledge Management (to appear, 2007)
12. Page, L., Brin, S., Motwani, R., Winograd, T.: The PageRank citation ranking: Bringing order to the web. Technical report, Stanford Digital Library Technologies Project (1998)

Parallelizing the Computation of PageRank

John Wicks and Amy Greenwald

Department of Computer Science
Brown University, Box 1910
Providence, RI 02912
{jwicks,amy}@cs.brown.edu)

Abstract. This paper presents a technique we call **ParaSolve** that exploits the sparsity structure of the web graph matrix to improve on the degree of parallelism in a number of distributed approaches for computating PageRank. Specifically, a typical algorithm (such as power iteration or GMRES) for solving the linear system corresponding to PageRank, call it **LinearSolve**, may be converted to a distributed algorithm, **Distrib(LinearSolve)**, by partitioning the problem and applying a standard technique (i.e., **Distrib**). By reducing the number of inter-partition multiplications, we may greatly increase the degree of parallelism, while achieving a similar degree of accuracy. This should lead to increasingly better performance as we utilize more processors. For example, using **GeoSolve** (a variant of Jacobi iteration) as our linear solver and the 2001 web graph from Stanford's WebBase project, on 12 processors **ParaSolve(GeoSolve)** outperforms **Distrib(GeoSolve)** by a factor of 1.4, while on 32 processors the performance ratio improves to 2.8.

1 Introduction

The first order, homogeneous, linear recurrence:

$$\boldsymbol{w}_{n+1} = A\boldsymbol{w}_n + \boldsymbol{b} \tag{1}$$

occurs naturally in various settings. When $\|A\|_1 < 1$, it is well-known that $\boldsymbol{w}_n = A^n\boldsymbol{w}_0 + \sum_{j=0}^{n-1} A^j\boldsymbol{b}$ and $\boldsymbol{w}_n \to \boldsymbol{w} \equiv \sum_{j=0}^{\infty} A^j\boldsymbol{b}$, the fixed-point of Equation 1, independent of $\boldsymbol{w}_0$.[1] In other words,

$$\boldsymbol{w} = A\boldsymbol{w} + \boldsymbol{b} \tag{2}$$

For example, computing PageRank [10] via power iteration leads to an instance of this recurrence. Specifically, given a web graph matrix, $M \geq 0$, with "normalized" columns (i.e., each column sums to 1), a (normalized) personalization vector, $\boldsymbol{v} \geq 0$, and a teleportation probability, ϵ, define the perturbed Markov matrix, $M_{\boldsymbol{v},\epsilon} = (1 - \epsilon)M + \epsilon\boldsymbol{v}\boldsymbol{J}^t$, where $\boldsymbol{J}_i = 1, \forall\, i$. Power iteration takes an arbitrary, normalized initial vector, $\boldsymbol{v}_0 \geq 0$, computes $\boldsymbol{r}_{n+1} = M_{\boldsymbol{v},\epsilon}\boldsymbol{r}_n$, with $\boldsymbol{r}_0 = \boldsymbol{v}_0$, and terminates when $\|\boldsymbol{r}_n - \boldsymbol{r}_{n-1}\|_1 < \delta$. Since $M_{\boldsymbol{v},\epsilon}$ and $\boldsymbol{v}_0$ are

---

[1] $\|A\|_1 \equiv \max_{\|\boldsymbol{w}\|_1=1} \|A\boldsymbol{w}\|_1$, with $\|\boldsymbol{w}\|_1 \equiv \max_i |\boldsymbol{w}_i|$.

A. Bonato and F.R.K. Chung (Eds.): WAW 2007, LNCS 4863, pp. 202–208, 2007.
© Springer-Verlag Berlin Heidelberg 2007

normalized, so is r_n, $\forall n$. This implies that $r_{n+1} = (1 - \epsilon)Mr_n + \epsilon v$, which is just Equation 1 with $A = (1 - \epsilon)M$ and $b = \epsilon v$. Therefore, r_n converges to $r = \epsilon \sum_{j=0}^{\infty} [(1 - \epsilon)M]^j v$ for any v_0. In particular, r is the unique positive, normalized eigenvector of $M_{v,\epsilon}$ with eigenvalue 1, which is the usual definition of the PageRank vector.

There have been many attempts to speed up the PageRank computation. Since Equation 2 is just a linear equation, there are a wide variety of algorithms for computing approximate solutions, including the Jacobi, Gauss-Seidel, Generalized Minimum Residual (GMRES) and Biconjugate Gradient (BiCG) methods [8, 6]. Although they vary in many respects, such algorithms tend to require the computationally intensive step of matrix-vector multiplication, Aw. One class of approaches involves partitioning A and distributing the partitions among several machines, so that A may be kept in main memory and the multiplication may be performed in a distributed fashion, using a standard technique described below.

By exploiting the sparsity structure of the web graph matrix, it should be possible, in principle, to perform the computation efficiently in parallel. Kamvar, et al. [7] have observed that, when the pages are properly ordered, this matrix is concentrated along the diagonal. Moreover, Kohlschütter et al. [8] noted that when pages are grouped according to "site" name (e.g., "brown.edu" or "http://www.cs.brown.edu/people/"), the matrix is almost block-diagonal, with blocks corresponding to sites. By merging site blocks, one obtains an almost block-diagonal matrix with relatively few large blocks.

Define A_0 to be the matrix consisting only of the diagonal blocks of A, and let $A_1 = A - A_0$. That is, A_0 consists (primarily) of links within any given site, while A_1 consists entirely of links between sites. Multiplication by $A = A_0 + A_1$ is effectively multiplication by A_0 plus multiplication by A_1. Since sites tend to have more internal than external links, A_1 is much more sparse than A_0, which is what we mean by "almost block-diagonal".

Multiplication by A_0 can be performed in parallel, since A_0 is block-diagonal. Moreover, since A_0w dominates A_1w, it may be worthwhile to perform A_0 multiplications more often than A_1 ones. Since, as we will see, multiplication by A_1 is the major bottle-neck in the distributed multiplication, this should lead to an increase in performance. In particular, *by reducing the number of A_1 multiplications, we should be able to increase the amount of computation done in parallel, without sacrificing accuracy.* This intuition leads to our technique, **ParaSolve**, which we believe will yield more efficient distributed algorithms to solve Equation 2, when A is almost block-diagonal, and hence to compute PageRank.

To illustrate, suppose for simplicity that w is partitioned into three segments, w_i, and A is partitioned into 9 corresponding blocks, $A_{i,j}$, $i, j = 0, \ldots, 2$. Three machines may then compute $\overline{w} = Aw$, in straightforward manner as follows. Machine j stores $A_{i,j}$ and w_j, computes $\hat{w}_{i,j} = A_{i,j}w_j$, $i = 0, \ldots, 2$, sends $\hat{w}_{i,j}$, $i \neq j$ to machine i, and accumulates the results, $\overline{w}_j = \hat{w}_{j,j} + \sum_{i \neq j} \hat{w}_{j,i}$. When A is the web graph, Kohlschütter et al. [8] found that the off-diagonal products, $\hat{w}_{i,j}$, $i \neq j$ contributing to A_1w are sparse, and noted that they may be transmitted at negligible cost over a 1Gb network.

Some have used this manner of distributed multiplication to solve Equation 2 in a distributed fashion [8, 6]. Each method used was, at heart, the result of applying this technique to a standard linear solution algorithm, call it **Linear-Solve**, to obtain a distributed version of the algorithm, **Distrib(LinearSolve)**. While those researchers referred to these methods as "parallel", this terminology is a bit misleading.

While, in theory, the multiplications, $A_{i,j}\boldsymbol{w}_j$, can be performed in time proportional to the number of non-zero entries in $A_{i,j}$, as the number of rows in the matrix grows, the cost of cache misses in storing the resulting vectors, $\hat{\boldsymbol{w}}_{i,j}$, becomes a significant factor. In the case of the Web, the dimension of A is sufficiently large for this to dominate so that *the cost of performing the multiplications on any one processor stays essentially constant*, even as the number of partitions is increased. In particular, the speed of computation does *not* increase proportional to the number of processors, as one would expect with a truly parallel algorithm. Moreover, distributed multiplication by A_1 is a synchronized operation, requiring a great deal of inter-process communication, while "parallel" suggests that the computations can be carried out (for the most part) independently.

In this paper, we will present an alternative technique, **ParaSolve**, to distribute any given algorithm, **LinearSolve**, to compute approximate solutions to linear systems. We claim that, when applied to almost block-diagonal systems, such as that for computing PageRank, the resulting distributed algorithm, **ParaSolve(LinearSolve)**, will be superior to the standard distributed version, **Distrib(LinearSolve)**, due to its higher degree of parallelism, decreased need for synchronization, reduced volume of interprocess communication, and, in some cases, reduced memory load.

We have currently applied **ParaSolve** to a variant of Jacobi iteration, **GeoSolve**, and present experiments to compare how **ParaSolve(GeoSolve)** behaves in practice on the 2001 web graph from Stanford's WebBase project. In particular, we show that, due to the reduced number of A_1 multiplications, **ParaSolve(GeoSolve)** scales well as the number of machines increases. For example, on 12 processors **ParaSolve(GeoSolve)** outperforms **Distrib(GeoSolve)** by a factor of 1.4, while on 32 processors the performance ratio improves to 2.8. We go on to indicate why, as the number of processors increases, this performance ratio can be expected to improve even more. We also discuss why this ratio may be even better for other linear solvers, such as GMRES.

2   ParaSolve

Our **ParaSolve** technique is closely related to Jacobi iteration. Ignoring the preconditioning of A by its diagonal entries, Jacobi iteration simply iterates Equation 1 to its fixed-point, $\boldsymbol{w} \equiv \sum_{j=0}^{\infty} A^j \boldsymbol{b} = [I - A]^{-1} \boldsymbol{b} \equiv A^{\infty} \boldsymbol{b}$. If we take $\boldsymbol{b}$ as $\boldsymbol{w}_0$ and iterate Equation 1, we obtain the partial sums of $A^{\infty}\boldsymbol{b}$, which converge if $\|A\|_1 < 1$. In general, any approximate linear solver for Equation 2 may be viewed as yielding an approximation to the operation of multiplying $\boldsymbol{b}$ by A^{∞}.

Now assume, as in the Introduction, that $A \geq 0$ has been decomposed into $A_0 \geq 0$ and $A_1 \geq 0$. Expanding out the powers of $[A_0 + A_1]^j$ in A^∞ yields a sum over words in A_i. Notice that $A_0^\infty = \sum_{j=0}^\infty A_0^j$ is the sum over arbitrary length words only in A_0. Grouping terms according to the number of A_1 factors gives $w = A^\infty b = \sum_{j=0}^\infty [A_0 + A_1]^j b = \sum_{j=0}^\infty \sum_{d \in \{0,1\}^j} \prod_{i=0}^{j-1} A_{d_i} b = [A_0^\infty + A_0^\infty A_1 A_0^\infty + \ldots] b$, i.e.:

$$w = \sum_{j=0}^\infty A_0^\infty [A_1 A_0^\infty]^j b \tag{3}$$

This is all legitimate, since we are dealing with absolutely convergent series.

Since multiplication by A_0^∞ may be approximated using a linear solver, Equation 3 suggests the following algorithm:

ParaSolve(A_0, A_1, b, δ)
   initialize α, $s_0 = b$, $t_0 = \textbf{LinearSolve}\,(A_0, s_0, \alpha\delta)$, and $w_0 = t_0$
   update $s_{n+1} = A_1 t_n$, $t_{n+1} = \textbf{LinearSolve}\,(A_0, s_{n+1}, \alpha\delta)$, and
   $w_{n+1} = w_n + t_{n+1}$
   terminate $\|t_n\|_1 < \delta$

When $A_0 = 0$, so we may take $t_n = s_n$, we obtain a linear solver, akin to Jacobi iteration, which we call **GeoSolve**.

When A is almost block-diagonal, with diagonal component A_0, then the update, $t_{n+1} = \textbf{LinearSolve}\,(A_0, s_n, \alpha\delta)$ may be performed in parallel, while $s_{n+1} = A_1 t_n$ may be distributed, as described in the Introduction, yielding the corresponding distributed algorithm, **ParaSolve(LinearSolve)**. Since A_1 is small, **ParaSolve** will converge after very few iterations.

Notice that, since any error in an approximation of t_n is decreased by its subsequent multiplication by A_1, **LinearSolve** may be computed to a proportionately larger error tolerance, $\alpha\delta$. For example, we could take $\alpha = \|A_1\|^{-1}$. Assuming that **LinearSolve** is continuous in its inputs, **ParaSolve(LinearSolve)** will converge to the solution of Equation 2. Moreover, since w_n converges, t_n converges to 0. In particular, since we need only compute t_n to within $\alpha\delta$, the cost of subsequent calls to **LinearSolve** decreases.

3   Experiments

To see how **ParaSolve** and **Distrib** compare in practice, we used the (decompressed) version of Stanford's web graph, based on a 2001 crawl as part of its WebBase project and **GeoSolve** (cf. Section 2) as our linear solver. The data provided by WebBase was filtered to eliminate invalid links and to normalize URLs and distributed courtesy of the WebGraph project [2]. Both algorithms were implemented in C++, using the Matrix Template Library (MTL) [4].

This graph has about 10^8 nodes and 10^9 links. We re-indexed the pages so that those within common sites (defined by primary host name) were contiguous. The (normalized) web graph matrix and ranking vector were partitioned, by merging sites, so that the number of links was approximately equal across machines. We used a uniform distribution personalization vector, v, i.e., $v_i = (\dim v)^{-1}$.

The following experiments used a cluster of Apple PowerPC G5 (3.0) XServes with dual 2GHz processors, 2G of RAM, 512 KB L2 cache per CPU, running OS X Server 10.4.7 connected by a 1Gb ethernet in the Brown Internet Lab [3]. There was only one other user on these machines, and he continuously ran a few background jobs with low CPU usage. Although these machines were not completely dedicated, all runs were performed under identical conditions. Timings are in CPU-seconds, excluding time to load the web graph and write the final results out to disk. Table 1 shows how the two algorithms performed as we varied the number of machines with δ fixed at 10^{-3}.

Table 1. ParaSolve vs. Distrib with GeoSolve

# of Machines	ParaSolve			Distrib		Ratio
	Time	A_0	A_1	Time	A	
12	581	5.9	26	831	26	1.4
16	405	4.2	18	685	21	1.7
20	**327**	**3.3**	**18**	**669**	**21**	**2.1**
24	276	2.7	19	653	20	2.4
28	250	2.3	19	649	20	2.6
32	230	2.1	19	650	20	2.8

We consider, in some detail a typical run of each algorithm with $k = 20$ machines. Each partition was roughly 6×10^6 dimensional and, on average, the number of non-zero entries in $A_{j,j}$, nnz $(A_{j,j}) \approx 5 \times 10^7$, while nnz $(A_{i,j}) \approx 10^5$ for $i \neq j$, which confirms our hypothesis that A_0 is much larger than A_1. Likewise, on average, the size of the component of t in the ith partition, nnz $(t_i) \approx 3 \times 10^5$, which confirms our sparsity assumption on the off-diagonal products. **Para-Solve** converged by $i = 3$ in 327 sec. Each **GeoSolve** call took, on average, 3.3 sec./mult., with 41 multiplications at $i = 0$, 24 multiplications at $i = 1$, 11 multiplications at $i = 2$, and 2 multiplications at $i = 3$. The remaining computational time was spent in performing the off-diagonal multiplications (i.e., multiplying by A_1), which took, on average, 18 sec. per iteration, for a total cost per machine of 54 sec., which was less than one sixth of the total time.

While **ParaSolve** effectively performed 78 multiplications by A_0 and 3 multiplications by A_1, by comparison, **Distrib** required 31 multiplications by A_0 and A_1 to converge. Since the cost of an A_1 multiplication is on the order of 10 times that of an A_0 multiplication, over 90% of the time in **Distrib** is spent multiplying by A_1. In contrast, **ParaSolve** spends only about 17% of its time performing off-diagonal multiplications. While the total number of multiplications is less to carry out **Distrib** as compared to **ParaSolve**, the cost of an off-diagonal multiplication is much higher, so that **Distrib** took 2.1 times as much CPU time.

As we varied the number of machines, both algorithms required roughly the same number of multiplications of each type (i.e., A_0 vs. A_1). But as the number of machines increases, the factor by which **ParaSolve** outperforms **Distrib** increases. This is because, as we pointed out in the Introduction, the the cost of an A_1 multiplications does not decrease as we increase the number of machines. In contrast, as the dimensions of the diagonal blocks decrease, the cost of A_0 multiplications tends to 0. Thus, asyptotically the performance ratio should approach the ratio of the respective number of A_1 multiplications, which in this case is $31/3 \approx 10$.

4   Future Work

While our experiments have shown that **ParaSolve** can improve a particular distributed algorithm for computing PageRank, much more work remains to be done. Timing experiments have shown that an alternative matrix package, PETSc [5], can perform sparse matrix multiplication on average 30 times faster than MTL. Moreover, it provides a wide variety of linear solvers in both single machine and distributed settings. By retooling our implementation with PETSc, we will be able to quickly evaluate **ParaSolve** when applied to other linear solvers, such as GMRES and Biconjugate Gradient.

In contrast to **GeoSolve** which is a stateless solver, GMRES is based on an iterative decomposition of A. Since PETSc saves the state of the decomposition between subsequent calls, successive calls to a GMRES solver should be quite fast. While Gleich et al. [6] have shown **Distrib(GMRES)** to work well on smaller web graphs, the additional state leads to an increasing and excessive memory load as the algorithm progresses. In contrast, the memory requirements of **ParaSolve(GMRES)** are much more modest, since GMRES is only applied locally to the diagonal blocks of A_0.

We should point out that, although he did not actually implement it as a distributed system, McSherry [9] has suggested an alternative scheme to reduce the amount of cross-machine multiplications. While it is not as general as our technique, it has the advantage that results may be easily updated as the web graph evolves. We need to explore the extent to which our technique accommodates the computation of such updates.

The time of each call to **LinearSolve** depends strongly on our choice of α. We used a very conservative estimate for our experiments. The L_1-norm is based on a worst-case analysis and so is inherently conservative. A better choice of α might follow from considering the *average* case. In general, further numerical analysis of our algorithm is necessary to improve performance and provide strong error bounds.

Acknowledgments

Thanks to the Laboratory for Web Algorithmics[1] for providing the web graph data, as well as Prof. Steven P. Reiss for the use of the Brown University Internet Laboratory[3] for our experiments.

References

[1] http://law.dsi.unimi.it/
[2] http://webgraph.dsi.unimi.it/
[3] http://www.cs.brown.edu/rooms/ilab/
[4] http://www.osl.iu.edu/research/mtl/
[5] Balay, S., Buschelman, K., Eijkhout, V., Gropp, W.D., Kaushik, D., Knepley, M.G., McInnes, L.C., Smith, B.F., Zhang, H.: PETSc users manual. Technical Report ANL-95/11 - Revision 2.1.5, Argonne National Laboratory (2004)
[6] Gleich, D., Zhukov, L., Berkhin, P.: Fast parallel pagerank: A linear system approach. In: WWW 2005: Proceedings of the 14th international conference on World Wide Web, ACM Press, New York (2005)
[7] Kamvar, S., Haveliwala, T., Manning, C., Golub, G.: Exploiting the block structure of the web for computing pagerank. Technical report, Stanford University Technical Report (2003)
[8] Kohlschütter, C., Chirita, P.-A., Nejdl, W.: Efficient parallel computation of pagerank. In: Lalmas, M., MacFarlane, A., Rüger, S., Tombros, A., Tsikrika, T., Yavlinsky, A. (eds.) Advances in Information Retrieval. LNCS, vol. 3936, pp. 241–252. Springer, Heidelberg (2006)
[9] McSherry, F.: A uniform approach to accelerated pagerank computation. In: WWW 2005: Proceedings of the 14th international conference on World Wide Web, pp. 575–582. ACM Press, New York (2005)
[10] Page, L., Brin, S., Motwani, R., Winograd, T.: The pagerank citation ranking: Bringing order to the web. Technical report, Stanford Digital Library Technologies Project (1998)

Giant Component and Connectivity in Geographical Threshold Graphs

Milan Bradonjić[1], Aric Hagberg[2], and Allon G. Percus[3,4]

[1] Department of Electrical Engineering, UCLA, Los Angeles, CA 90095
milan@ee.ucla.edu
[2] Mathematical Modeling and Analysis, Theoretical Division,
Los Alamos National Laboratory, Los Alamos, NM 87545
hagberg@lanl.gov
[3] Department of Mathematics, UCLA, Los Angeles, CA 90095
[4] Information Sciences Group, Los Alamos National Laboratory, Los Alamos, NM 87545
percus@ipam.ucla.edu

Abstract. The geographical threshold graph model is a random graph model with nodes distributed in a Euclidean space and edges assigned through a function of distance and node weights. We study this model and give conditions for the absence and existence of the giant component, as well as for connectivity.

Keywords: random graph, geographical threshold graph, giant component, connectivity.

1 Introduction

Large networks such as the Internet, World Wide Web, phone call graphs, infections disease contacts, and financial transactions have provided new challenges for modeling and analysis [1]. For example, Web graphs may have billions of nodes and edges, which implies that the analysis on these graphs, i.e., processing and extracting information on these large sets of data, is "hard" [2]. Extensive theoretical and experimental research has been done in web-graph modeling. Early measurements suggested that the Internet exhibits a power-law degree distribution [3] and that the web graph also follows a power-law distribution in in- and out-degree of links [4]. Modeling approaches using random graphs have attempted to capture both the structure and dynamics of the web graph [5,6,7,8,9].

In this short paper we study geographical threshold graphs (GTGs), a static model for networks that includes both geometric information and node weight information. The motivation for analyzing this model is that many real networks need to be studied by using a "richer" stochastic model (which in this case includes both a distance between nodes and weights on the nodes). This model has already been applied in the study of wireless ad hoc networks for systems where the wireless nodes have different capabilities [10]. The weights, in this case, represent power or bandwidth resources available to wireless nodes in the network. By varying the weights in a GTG model, properties such as the diameter or degree distribution can be tuned. Other possible applications of GTGs are in epidemic modeling, where the weights might represent susceptibility to

A. Bonato and F.R.K. Chung (Eds.): WAW 2007, LNCS 4863, pp. 209–216, 2007.
© Springer-Verlag Berlin Heidelberg 2007

infection, or other social networks where the weights may be related to attractiveness or other individual characteristic.

2 Geographical Threshold Graph Model

In addition to unstructured random graphs [11,12], recent research has focused on random geometric graphs (RGG) where edges are created according to a distance between nodes [13], and threshold graphs [14,15] with edges created according to a function of node weights. Geographical threshold graphs, which combine aspects of RGG and threshold graphs, have only recently begun to receive attention [16,10].

The GTG model is constructed from of a set of n nodes placed independently in $\mathbf{R}^d$ according to a Poisson point-wise process. A non-negative weight w_i, taken randomly and independently from a probability distribution function $f(w) : \mathbf{R}_0^+ \to \mathbf{R}_0^+$, is assigned to each node v_i for $i \in [n]$. Let $F(x) = \int_0^x f(w)dw$ be the cumulative density function. For two nodes i and j at distance r, the edge (i, j) exists if and only if the following connectivity relation is satisfied:

$$G(w_i, w_j)h(r) \geq \theta_n, \tag{1}$$

where θ_n is a given threshold parameter that depends on the size of the network. The function $h(r)$ is assumed to be decreasing in r. We use $h(r) = r^{-\alpha}$, for some positive α, which is typical for e.g., the path-loss model in wireless networks [10]. The interaction strength between nodes $G(w_i, w_j)$ is usually taken to be symmetric and either multiplicatively or additively separable, i.e., in the form of $G(w_i, w_j) = g(w_i)g(w_j)$ or $G(w_i, w_j) = g(w_i) + g(w_j)$.

Some basic results have already been shown. For the case of uniformly distributed nodes over a unit space it has been shown [16,10] that the expected degree of a node with weight w is

$$\mathrm{E}[k(w)] = \frac{n\pi^{d/2}}{\Gamma(d/2+1)} \int_{w'} f(w')\left(h^{-1}(\theta_n/G(w, w'))\right)^d dw', \tag{2}$$

where h^{-1} is the inverse of h. The degree distribution has been studied for specific weight distribution functions $f(w)$ [16]. In both the multiplicative and additive case of $G(w, w')$, questions of diameter, connectivity, and topology control have been addressed [10].

Here we restrict ourselves to the case of $g(w) = w$, $\alpha = 2$, and nodes distributed uniformly over a two-dimensional space. For analytical simplicity we take the space to be a unit torus. We concentrate on the analysis of the additive model, i.e., when the connectivity relation is given by

$$\frac{w_i + w_j}{r^2} \geq \theta_n. \tag{3}$$

Our techniques may be generalized to other cases in a straightforward manner. Our contribution in this short paper is to provide the first bounds on θ_n for the emergence of the giant component, and for connectivity.

3   Giant Component in GTG

Definition 1 (Giant Component). *The giant component is a connected component with size $\Theta(n)$.*

In this section we analyze the conditions for the existence of the giant component, giving bounds on the threshold parameter value θ_n where it first appears. For $\theta_n = cn$, we specify positive constants $c' > c''$ and prove that whp, if $c > c'$ the giant component does not exist whereas if $c < c''$ the giant component exists.

3.1   Absence of Giant Component

Lemma 1. *Let $\theta_n = cn$ for $c > c'$, where $c' = 2\pi E[w]$. Then whp there is no giant component in GTG.*

Proof. We use an approach similar to one given in [17]. Divide the nodes into three classes: alive, dead and neutral. Denote the number of alive nodes as Y_i. The algorithm works as follows. At time $t = 0$, designate one node (picked u.a.r.) as being alive and all others as neutral. Now, at each subsequent time step t, pick a node v_t u.a.r. from among those that are alive, and then consider all neutral nodes connected to v_t. Denote the number of these nodes as Z_t. Change these nodes from neutral to alive, and change v_t itself from alive to dead. The random variables Y_i, Z_i satisfy the following recursion relation: $Y_0 = 1$ and $Y_t = Y_{t-1} + Z_t - 1$, for $t \geq 1$. The number of alive nodes satisfies

$$Y_t - 1 = \sum_{k=1}^{t} Z_k - t. \tag{4}$$

At a time step k, let d_k be the degree of node v_k. Since Z_k only includes the neutral nodes connected to v_k,

$$Z_k \leq d_k. \tag{5}$$

Now let T be the largest t such that $Y_t > 0$. Then T is the size of the component containing v_0, and the giant component exists if and only if $T = \Theta(n)$ with some nonvanishing probability. The variable T satisfies the following relation

$$\Pr[T \geq t] = \Pr[Y_t > 0] = \Pr[Y_t \geq 1] = \Pr[\sum_{k=1}^{t} Z_k \geq t] \leq \Pr[\sum_{k=1}^{t} d_k \geq t]. \tag{6}$$

Consider the threshold $\theta_n = cn$ for some $c > 0$. It is shown in the Appendix that for a node v_k with random weight w_k, the vertex degree distribution is Poisson: $d(w_k) \sim Po(a(w_k + \mu))$, where $a = n\pi/\theta_n$ and $\mu = E[w]$. Since the sum of independent random Poisson variables is a Poisson random variable,

$$\Pr\left[\sum_{k=1}^{t} d_k \geq t\right] = \Pr\left[Po(a \sum_{k=1}^{t} (w_k + \mu)) \geq t\right]. \tag{7}$$

We now use the following inequality. For any $\varepsilon \in (0,1)$,

$$\Pr\left[Po(a\sum(w_k+\mu)) \geq t\right] \leq \Pr\left[Po(a\sum(w_k+\mu)) \geq t \mid \sum w_k \in (1\pm\varepsilon)t\mu\right]$$
$$+ \Pr\left[\sum w_k \notin (1\pm\varepsilon)t\mu\right].$$

By the central limit theorem, for $t \to \infty$, the sum $(\sum w_k - t\mu)/(\sqrt{t}\sigma)$ tends to the normal distribution $N(0,1)$. That is,

$$\Pr\left[\sum w_k \notin (1\pm\varepsilon)t\mu\right] = \Pr\left[\frac{\sum w_k - t\mu}{\sqrt{t}\sigma} \notin (-\varepsilon,\varepsilon)\sqrt{t}\frac{\mu}{\sigma}\right] \to 0. \qquad (8)$$

Finally, we use the concentration on the Poisson random variable [13]. Define $\lambda = \mathrm{E}[a\sum(w_k+\mu)] = 2at\mu$. Given any $\varepsilon_0 \in (0,1)$, for $t \to \infty$, i.e., $\lambda \to +\infty$, it follows that

$$\Pr[Po(\lambda) \notin (1\pm\varepsilon_0)\lambda] \leq e^{-\lambda H(1-\varepsilon_0)} + e^{-\lambda H(1+\varepsilon_0)} \to 0, \qquad (9)$$

where the function $H(x) = 1 - x + x\ln x$, for $x > 0$. It is now sufficient to choose a small enough that $t > 2at\mu(1+\varepsilon_0)$ for some positive constant ε_0. This is equivalent to $1 > 2a\mu$, i.e., $c > 2\pi\mu$. It follows that $\Pr[Po(a\sum(w_i+\mu)) \geq t] = o(1)$ for $t = \Theta(n)$, which completes the proof.

3.2   Existence of Giant Component

Lemma 2. *Let $\theta_n = cn$ for $c < c'' = \sup_{\alpha\in(0,1)} \alpha F^{-1}(1-\alpha)/\lambda_c$, where $\pi\lambda_c \approx 4.52$ is the mean degree at which the giant component first appears in Random Geometric Graphs (RGG) [13]. Then whp the giant component exists in GTG.*

Proof. For any constant $\alpha \in (0,1)$, we prove that whp there are αn "high-weighted" nodes, all with weights greater than or equal to some s_n; we state s_n later. Let X_i be the indicator of the event $W_i \geq s_n$. Then $\Pr[X_i = 1] = 1 - F(s) =: q$. Let $X = \sum_{i=1}^{n} X_i$ be the number of high-weighted nodes. Using the Chernoff bound $\Pr[X \leq (1-\delta)\mathrm{E}[X]] \leq \exp(-\mathrm{E}[X]\delta^2/2)$, with $\delta = 1 - \alpha/q$,

$$\Pr[X \leq \alpha n] = \Pr[X \leq (1-\delta)\mathrm{E}[X]] \leq \exp\left(-n(q-\alpha)^2/(2q)\right) = n^{-\beta} \qquad (10)$$

for some constant $\beta > 1$ satisfying $(q-\alpha)^2 = 2q\beta\ln n/n$. Solving that quadratic equation in q gives $q = \alpha + \Theta(\ln n/n)$, so $F(s_n) = 1 - q = 1 - \alpha - \Theta(\ln n/n)$. For any $\varepsilon > 0$ and n sufficiently large the following is satisfied

$$F^{-1}(1-\alpha) \geq s_n \geq F^{-1}(1-\alpha-\varepsilon). \qquad (11)$$

Thus, let us define the sequence s_n by its limit

$$s_n \to F^{-1}(1-\alpha) = \Theta(1). \qquad (12)$$

Now we consider the set of αn high-weighted nodes. For each such node v_i with weight w_i, define its characteristic radius to be

$$r_t^2(w_i) = w_i/\theta_n. \qquad (13)$$

Then it follows that any other high-weighted node v_j within this radius is connected to v_i, since the connectivity relation is satisfied:

$$(w_i + w_j)/r^2 \geq w_i/r_t^2 = \theta_n. \tag{14}$$

Let $\theta_n = cn$, where $c < \alpha F^{-1}(1-\alpha)/\lambda_c$. For the radius r_t, whp it follows

$$r_t^2(w_i) = \frac{w_i}{\theta_n} \geq \frac{s_n}{\theta_n} > \frac{\lambda_c}{\alpha n}. \tag{15}$$

Let us therefore consider small circles, with a fixed radius r_0 s.t. $\sqrt{s_n/\theta_n} > r_0 > \sqrt{\lambda_c/(\alpha n)}$, around each of these αn nodes. A subgraph of this must be a RGG with mean degree $> \lambda_c$, which whp contains a giant component. Since its size is $\Theta(\alpha n) = \Theta(n)$, it is a giant component of the GTG too. We may optimize the bound by taking the supremum of c over $\alpha \in (0, 1)$, and the lemma follows.

4   Connectivity in GTG

Definition 2 (Connectivity). *The graph on n vertices is connected if the largest component has size n.*

In this section we analyze sufficient conditions for the entire graph to be connected. We consider the connectivity threshold $\theta_n = cn/\ln n$ and specify a bound on c.

Lemma 3. *Let $\theta_n = cn/\ln n$ for $c < \sup_{\alpha \in (0,1)} \alpha F^{-1}(1-\alpha)/4$. Then the GTG is connected whp.*

Proof. The proof is divided into two parts. In the first part, we prove that a constant fraction of nodes αn are connected. In the second part we prove that the rest of the $(1-\alpha)n$ nodes are connected to the first set of αn nodes.

First part: Invoking the proof of the appearance of the giant component, there are αn nodes all with weights $\geq s_n \to F^{-1}(1-\alpha) = \Theta(1)$.

Let $\theta_n = cn/\ln n$, where $c < \alpha F^{-1}(1-\alpha)\pi$. Analogously to r_t, define the connectivity radius r_c

$$r_c^2(w_i) = \frac{w_i}{\theta_n} \geq \frac{s_n}{\theta_n} > \frac{\ln n}{\alpha \pi n}. \tag{16}$$

Similarly to Lemma 2 let us consider small circles around each of these αn nodes, and consider these nodes as a RGG. It is known that $r_n = \sqrt{\ln n/(\pi n)}$ is the connectivity threshold in RGG [18]. The connectivity of RGG implies the connectivity of these αn nodes in our GTG.

Second part: Color the αn high-weighted nodes blue, and the remaining $(1-\alpha)n$ nodes red. Now let us tile our space into $n/(c_0 \ln n)$ squares of size $c_0 \ln n/n$. We state c_0 later. Consider any square S_i, and let B_i be the number of blue nodes in S_i. In expectation there are $E[B_i] = \alpha c_0 \ln n$ blue nodes in each square. From the Chernoff bound it follows

$$\Pr[B_i \geq (1-\delta)\alpha c_0 \ln n] \geq 1 - n^{-\alpha c_0 \delta^2/2}. \tag{17}$$

Let us consider one red node r. The node r belongs to some square S_i. Let M_r be the event that the red node r is connected to some blue node $b \in S_i$. Let the weights of r, b be w_r, w_b, respectively. The probability of the complement of M_r, conditioned on there being at least one blue node in S_i, is given by

$$\Pr[M_r^c | B_i \geq 1] = \Pr[w_r + w_b \leq r^2 \theta_n] \leq \Pr[w_r + w_b \leq 2c_0 \frac{\ln n}{n} c \frac{n}{\ln n}]$$
$$= \Pr[w_r + w_b \leq 2c_0 c]. \tag{18}$$

As long as $F^{-1}(1-\alpha) > 2c_0 c$, $w_b > 2c_0 c$ and hence $\Pr[M_r^c | B_i \geq 1] = 0$. For large enough n it must hold that $(1-\delta)\alpha c_0 \ln n > 1$, and so from Eq. (17),

$$\Pr[M_r^c] \leq \Pr[M_r^c | B_i \geq (1-\delta)\alpha c_0 \ln n] + \Pr[B_i < (1-\delta)\alpha c_0 \ln n]$$
$$\leq 0 + n^{-\alpha c_0 \delta^2 / 2}. \tag{19}$$

If $\alpha c_0 \delta^2 / 2 \geq 1 + \varepsilon$ for some $\varepsilon > 0$, then by the union bound,

$$\Pr[\bigcup_r M_r^c] \leq \sum_r \Pr[M_r^c] \leq (1-\alpha)n n^{-(1+\varepsilon)} = (1-\alpha)n^{-\varepsilon}. \tag{20}$$

Finally, the probability that all red nodes are connected to the set of blue nodes is given by the following relation

$$\Pr[\bigcap_r M_r] = 1 - \Pr[\bigcup_r M_r^c] \geq 1 - (1-\alpha)n^{-\varepsilon} \to 1. \tag{21}$$

The requirements we have imposed on constants so far are: $c < \alpha F^{-1}(1-\alpha)\pi$, $c < F^{-1}(1-\alpha)/(2c_0)$ and $\alpha c_0 \geq 2(1+\varepsilon)/\delta^2$. These conditions combine to give

$$c < \alpha F^{-1}(1-\alpha) \min(\pi, \frac{\delta^2}{4(1+\varepsilon)}). \tag{22}$$

Since $\alpha \in (0,1)$, $\delta \in (0,1)$ and $\varepsilon > 0$ are arbitrary, we obtain

$$c < \sup_{\alpha \in (0,1)} \alpha F^{-1}(1-\alpha)/4. \tag{23}$$

5   Discussion

The GTG model is a versatile one and can be used not only for the generation and analysis of web-graphs or large complex networks, but more generally for relation graphs in a large data set. If the data have a metric and can be mapped to nodes in Euclidean space, much of the foregoing analysis applies: one may hope to control structural properties of the data set by studying it as a GTG.

Furthermore, while we considered the GTG model as a static structure, the set of weights in the model could vary in time. This would introduce dynamics, as might be appropriate for particular applications such as wireless networking.

Acknowledgements

Part of this work was funded by the Department of Energy at Los Alamos National Laboratory under contract DE-AC52-06NA25396 through the Laboratory Directed Research and Development Program.

References

1. Bonato, A.: A survey of models of the web graph. In: López-Ortiz, A., Hamel, A.M. (eds.) CAAN 2004. LNCS, vol. 3405, pp. 159–172. Springer, Heidelberg (2005)
2. Abello, J., Pardalos, P.M., Resende, M.G.C. (eds.): Handbook of massive data sets. Kluwer Academic Publishers, Norwell, MA (2002)
3. Faloutsos, M., Faloutsos, P., Faloutsos, C.: On power-law relationships of the internet topology. In: SIGCOMM 1999: Proceedings of the conference on Applications, technologies, architectures, and protocols for computer communication, pp. 251–262. ACM Press, New York (1999)
4. Kleinberg, J.M., Kumar, R., Raghavan, P., Rajagopalan, S., Tomkins, A.S.: The web as a graph: Measurements, models, and methods. In: Asano, T., Imai, H., Lee, D.T., Nakano, S.-i., Tokuyama, T. (eds.) COCOON 1999. LNCS, vol. 1627, pp. 1–17. Springer, Heidelberg (1999)
5. Kumar, R., Raghavan, P., Rajagopalan, S., Sivakumar, D., Tomkins, A., Upfal, E.: Stochastic models for the web graph. In: FOCS 2000: Proceedings of the 41st Annual Symposium on Foundations of Computer Science, Washington, DC, USA, p. 57. IEEE Computer Society, Los Alamitos (2000)
6. Barabási, A.L., Albert, R.: Emergence of scaling in random networks. Science 286, 509–512 (1999)
7. Aiello, W., Chung, F., Lu, L.: A random graph model for massive graphs. In: STOC 2000: Proceedings of the thirty-second annual ACM symposium on Theory of computing, pp. 171–180. ACM Press, New York (2000)
8. Bollobás, B., Riordan, O., Spencer, J., Tusnády, G.: The degree sequence of a scale-free random graph process. Random Struct. Algorithms 18, 279–290 (2001)
9. Cooper, C., Frieze, A.M.: A general model of undirected Web graphs. In: Meyer auf der Heide, F. (ed.) ESA 2001. LNCS, vol. 2161, pp. 500–511. Springer, Heidelberg (2001)
10. Bradonjić, M., Kong, J.: Wireless ad hoc networks with tunable topology. In: Proceedings of the 45th Annual Allerton Conference on Communication, Control and Computing (to appear, 2007)
11. Erdős, P., Rényi, A.: On random graphs. Publ. Math. Inst. Hungar. Acad. Sci. (1959)
12. Erdős, P., Rényi, A.: On the evolution of random graphs. Publ. Math. Inst. Hungar. Acad. Sci. (1960)
13. Penrose, M.D.: Random Geometric Graphs. Oxford University Press, Oxford (2003)
14. Mahadev, N.V.R., Peled, U.N.: Threshold graphs and related topics. Annals of discrete mathematics, vol. 56. Elsevier, New York (1995)
15. Hagberg, A., Swart, P.J., Schult, D.A.: Designing threshold networks with given structural and dynamical properties. Phys. Rev. E 74, 056116 (2006)
16. Masuda, N., Miwa, H., Konno, N.: Geographical threshold graphs with small-world and scale-free properties. Physical Review E 71, 036108 (2005)
17. Alon, N., Spencer, J.H.: The probabilistic method, 2nd edn. John Wiley & Sons, Inc., New York (2000)
18. Gupta, P., Kumar, P.R.: Critical power for asymptotic connectivity. In: Proceedings of the 37th IEEE Conference on Decision and Control, vol. 1, pp. 1106–1110 (1998)

Appendix

Degree Distribution

The nodes are placed into the unit torus. W.l.o.g. let us consider the degree of the node v_1. Let the weight vector be $\mathbf{w}$. Let the position vector of the nodes be $\mathbf{x}$. It is straightforward to show that the probability of v_1 having degree k, given weights $\mathbf{w}$, is

$$\Pr[d_1 = k|\mathbf{w}] = \binom{n-1}{k} \prod_{i=2}^{k+1} \text{Area}(B(x_i, r_{i1})) \prod_{j=k+2}^{n} (1 - \text{Area}(B(x_j, r_{j1}))), \qquad (24)$$

where $\text{Area}(B(x_i, r_{i1}))$ is the area of the ball at center x_i with radius r_{i1}, and due to (3) the radii are given by

$$r_{i1} = \sqrt{\frac{w_1 + w_i}{\theta_n}} \qquad (25)$$

for $i = 2, \dots, n$. After marginalization, it follows

$$\begin{aligned}
\Pr[d_1 = k|w_1] &= \left(\prod_{i=2}^{n} \int_{w_i} dw_i f(w_i) \right) \Pr[d_1 = k|\mathbf{w}] \\
&= \binom{n-1}{k} \left(\int_w dw f(w) \frac{(w_1 + w)\pi}{\theta_n} \right)^k \left(1 - \int_w dw f(w) \frac{(w_1 + w)\pi}{\theta_n} \right)^{n-1-k} \\
&= \binom{n-1}{k} \left(\frac{(w_1 + \mu)\pi}{\theta_n} \right)^k \left(1 - \frac{(w_1 + \mu)\pi}{\theta_n} \right)^{n-1-k} \\
&\to e^{-\lambda} \frac{\lambda^k}{k!},
\end{aligned}$$

where

$$\lambda = (w_1 + \mu)n\pi/\theta_n. \qquad (26)$$

That is, the degree distribution of a node with weight w, in the limit follows the Poisson distribution

$$d(k|w) \sim Po((w + \mu)n\pi/\theta_n). \qquad (27)$$

Author Index

Lecture Notes in Computer Science

Sublibrary 1: Theoretical Computer Science and General Issues

For information about Vols. 1– 4497
please contact your bookseller or Springer